AF737232

Semiconductors and Nanostructures

Semiconductors and Nanostructures: An Introduction provides a rigorous yet accessible introduction to semiconductor physics. Bridging the gap between introductory quantum mechanics and advanced semiconductor courses, it offers a compact yet comprehensive overview that empowers students to understand and engage with both fundamental concepts and real-world applications.

This textbook is an indispensable resource for undergraduate and graduate students in physics and engineering, particularly those preparing for advanced studies or careers in semiconductor technology. It is a practical, pedagogically refined resource for both coursework and self-study.

Key Features:

- Comprehensive yet concise coverage of core principles and advanced topics
- Systematic and didactic presentation of the needed concepts in statistical physics (such as the "Fermi level")
- Clear explanations of challenging concepts, supported by illustrative examples
- Research-connected problems linking theory to real-world applications
- Deeper focus on semiconductor defects than typical introductory texts

Rui César Vilão is Associate Professor at the Department of Physics of the University of Coimbra, as well as researcher of the Center of Physics of the University of Coimbra. He is a researcher in Condensed Matter Physics, working fundamentally in the investigation of the effects of the hydrogen impurity and of point defects in the properties of semiconductors and insulators. He is a co-author of over 70 research articles, essentially in works using positive muons as local probes of semiconductors and insulators, using the technique known as muon spin spectroscopy (μSR).

Textbook Series in Physical Sciences

This textbook series offers pedagogical resources for the physical sciences. It publishes high-quality, high-impact texts to improve understanding of fundamental and cutting edge topics, as well as to facilitate instruction. The authors are encouraged to incorporate numerous problems and worked examples, as well as making available solutions manuals for undergraduate and graduate level course adoptions. The format makes these texts useful as professional self-study and refresher guides as well. Subject areas covered in this series include condensed matter physics, quantum sciences, atomic, molecular, and plasma physics, energy science, nanoscience, spectroscopy, mathematical physics, geophysics, environmental physics, and so on, in terms of both theory and experiment.

Understanding Nanomaterials
Malkiat S. Johal

Concise Optics: Concepts, Examples, and Problems
Ajawad I. Haija, M. Z. Numan, W. Larry Freeman

A Mathematica Primer for Physicists
Jim Napolitano

Understanding Nanomaterials, Second Edition
Malkiat S. Johal, Lewis E. Johnson

Physics for Technology, Second Edition: With Applications in Industrial Control Electronics
Daniel H. Nichols

Time-Resolved Spectroscopy: An Experimental Perspective
Thomas Weinacht; Brett J. Pearson

No-Frills Physics: A Concise Study Guide for Algebra-Based Physics
Matthew D. McCluskey

Quantum Principles and Particles, Second Edition
Walter Wilcox

Electronic Conduction: Classical and Quantum Theory to Nanoelectronic Devices
John P. Xanthakis

Physics: A Focused Introduction
Colin Campbell

Quantum Many-Body Physics: A Path Integral Approach
Luca Salasnich

Semiconductors and Nanostructures: An Introduction
Rui Vilão

For more information about this series, please visit:
[www.crcpress.com/Textbook-Series-in-Physical-Sciences/book-series/TPHYSCI]

Semiconductors and Nanostructures

An Introduction

Rui César Vilão

CRC Press
Taylor & Francis Group
Boca Raton London New York

CRC Press is an imprint of the
Taylor & Francis Group, an **informa** business

Designed cover image: Shutterstock

First edition published 2026
by CRC Press
2385 NW Executive Center Drive, Suite 320, Boca Raton FL 33431

and by CRC Press
4 Park Square, Milton Park, Abingdon, Oxon, OX14 4RN

CRC Press is an imprint of Taylor & Francis Group, LLC

© 2026 Rui César Vilão

Reasonable efforts have been made to publish reliable data and information, but the author and publisher cannot assume responsibility for the validity of all materials or the consequences of their use. The authors and publishers have attempted to trace the copyright holders of all material reproduced in this publication and apologize to copyright holders if permission to publish in this form has not been obtained. If any copyright material has not been acknowledged please write and let us know so we may rectify in any future reprint.

Except as permitted under U.S. Copyright Law, no part of this book may be reprinted, reproduced, transmitted, or utilized in any form by any electronic, mechanical, or other means, now known or hereafter invented, including photocopying, microfilming, and recording, or in any information storage or retrieval system, without written permission from the publishers.

For permission to photocopy or use material electronically from this work, access www.copyright.com or contact the Copyright Clearance Center, Inc. (CCC), 222 Rosewood Drive, Danvers, MA 01923, 978-750-8400. For works that are not available on CCC please contact mpkbookspermissions@tandf.co.uk

For Product Safety Concerns and Information please contact our EU representative GPSR@taylorandfrancis.com. Taylor & Francis Verlag GmbH, Kaufingerstraße 24, 80331 München, Germany.

Trademark notice: Product or corporate names may be trademarks or registered trademarks and are used only for identification and explanation without intent to infringe.

ISBN: 978-1-032-96051-7 (hbk)
ISBN: 978-1-032-96052-4 (pbk)
ISBN: 978-1-003-58773-6 (ebk)

DOI: 10.1201/9781003587736

Typeset in CMR10 font
by KnowledgeWorks Global Ltd.

Publisher's note: This book has been prepared from camera-ready copy provided by the authors.

To Susana
To António and to Maria Beatriz

In memoriam
Maria Margarida Ramalho Ribeiro da Costa

Contents

Preface

This introductory textbook on semiconductor materials and nanostructures was developed as part of the teaching of this subject in the Master's program in Physics and the Master's program in Engineering Physics at the Faculty of Science and Technology of the University of Coimbra (FCTUC). The course syllabus proposes an introduction to semiconductor physics that not only allows the student to familiarize with the basic concepts necessary for an understanding of technological applications, but also reveals the immense fundamental physics hidden within them (in particular the quantum effects associated with nanostructures). This book seeks to respond to the need to balance the aspects linked to fundamental physics with those linked to applications, in order to give an overview of the richness of this topic. The coverage is not intended to be exhaustive, but rather sufficiently complete to allow this book to be used as a safe starting point for further developments. An attempt has been made to develop in additional depth some sub-topics linked to contemporary research, with an advantage for the critical training of students at this stage before they begin their master's dissertation. This link is reinforced by the inclusion of some problems based on more recent research results.

The introductory and specialized literature in this subject is quite abundant, but not easy to use in order to provide a broad, but solid overview of the main aspects of semiconductor physics in a single semester. This book tries to bridge this gap, in order to lay solid and clear foundations of the main results in semiconductor physics, giving the reader the confidence to move forward. After going through this book, the reader should have the essential physics concepts allowing him/her to venture into any specialized topic that he/she wishes in semiconductor physics and technology. The author is tributary to a couple of reference books in the field, which are indicated in the bibliography. Deep thanks are due to the colleagues in the Physics Department at FCTUC who have taught this subject, José António Paixão, Manuela Ramos Silva and Fernando Nogueira, for all their work, support and sharing over the years, as well as for the stimulating discussions from which the author has benefited so much. This work could also be enriched with some problems that they generously provided. A special word of thanks and gratitude is due to the late Prof. Margarida Costa, with whom the author had the privilege of starting to teach this subject, learning that scientific rigor and excellence can and should go side by side with grace, empathy and humanism. João Campos Gil, Helena Vieira Alberto and Alois Weidinger have accompanied the author throughout his scientific career, centered on semiconductor physics: this work would never have been dreamt up without their friendship and knowledge, which they always share so generously.

The book was first published in Portuguese at the Imprensa da Universidade de Coimbra (IUC – Coimbra University Press). Deep thanks are due to the IUC team for all their work in the Portuguese version. The director of IUC, Prof. Carlota Simões, is also deeply acknowledged for the encouragement and support of the publication of the current English version. A special word of thanks is due to the team of the CRC Press, who welcomed this project and helped the author in all stages, since the initial submission to the final printed version.

Finally, a last word for the students of Semiconductors and Nanostructures who, over the last few years, have been polishing these pages with their curiosity, commitment and enthusiasm.

1

Introduction and fundamentals

1.1 Introduction to semiconducting materials

Semiconductors are among the materials with the biggest impact on everyday life, on which much of modern technology is based, from transistors to LEDs and photovoltaic panels. It is therefore an area of heavy investment in both fundamental and applied research, in multidisciplinary teams of physicists, chemists, materials scientists and engineers.

From the perspective of the contemporary physicist, the physics of semiconductors is perhaps one of the best "visiting cards" of the discipline, linking in an exemplary way fundamental contributions (both theoretical and experimental) with the applied side and the transfer of knowledge and technology. This is eloquently summarized in the contributions that have been awarded the Nobel Prize in Physics since the growth of the discipline in the second half of the 20th century: transistor (1956), quantum electronics (1964), super lattices (1973), disordered systems (1977), quantum Hall effect (1985), fractional quantum Hall effect (1998), integrated circuit and heterostructures (2000), CCD camera (2009), graphene (2010) and blue LEDs (2014).

1.2 Electrons in solids

In this chapter, we present the essential ideas about the electronic structure of solids, which are needed for the presentation of the fundamentals of semiconductor physics. This is not intended to be an exhaustive treatment, which is beyond the scope and objectives of this book, but rather a summary of the essential points. We follow the traditional presentation, moving from Drude's[1] relaxation time model to Sommerfeld's[2] quantum model and then on to an introduction to band theory in solids.

[1]Paul Drude (1863-1906).
[2]Arnold Sommerfeld (1868-1951).

DOI: 10.1201/9781003587736-1

1.2.1 The relaxation time model

The Drude model is a classical model for electrical conduction in solids, which represented the first attempt to quantitatively describe these phenomena. It builds on the classical understanding of the properties of electrical conductors, based on the notion that free electrons (conduction electrons) must be present, and applies the ideas and methods of the classical kinetic theory of gases to these electrons. These concepts are supplemented with the concept of relaxation time, in order to supply a mechanism for the existence of resistivity. Despite the limitations of a classical model in the description of the behavior of the electron gas, the relaxation time model nevertheless is kept in more refined quantum models, so a brief review is needed.

The model is based on the idea that when an electric field is applied to a conducting material its electrons move while colliding in some way within the solid. The electrons are thus accelerated by the electric field, but after an average time τ (called *relaxation time*) they collide, and the average drift velocity of the electrons after the collision is zero. This approximation makes it possible, as shown below, to describe collisions as a viscous force, which is responsible for Ohm law.[3]

When electrons move with a drift velocity $\mathbf{v}_{\text{drift}}$, the corresponding current density $\mathbf{J}$ is:

$$\mathbf{J} = -ne\mathbf{v}_{\text{drift}} = -\frac{ne}{m_e}\mathbf{p} \tag{1.1}$$

where $\rho = -n\,e$ is the electron density of charge (n being the concentration of electrons and e the absolute value of their charge) and $\mathbf{p} = m_e\mathbf{v}_{\text{drift}}$ is the electronic linear momentum associated to the corresponding drift (m_e being the mass of the electron). Since τ is the average time between collisions, the probability of a collision in the time interval δt is $\delta t/\tau$. The probability that no collision occurs in the time interval δt is thus $1 - \delta t/\tau$. In a model of random collisions, electrons that collide in the time interval δt do not contribute with linear momentum after the collision, since their average drift velocity after the collision is zero. On the other hand, electrons that do not collide in the time interval δt are accelerated by the electric force $\mathbf{f}$, resulting in a variation of linear momentum $\delta\mathbf{p} = \mathbf{f}\delta t$. The linear momentum at the instant $t + \delta t$ thus results, on average:

$$\mathbf{p}(t + \delta t) = (1 - \delta t/\tau)\,[\mathbf{p}(t) + \mathbf{f}(t)\delta t] \tag{1.2}$$

Neglecting the contribution of the second-order terms in δt, we get:

$$\frac{d\mathbf{p}}{dt} = -\frac{\mathbf{p}}{\tau} + \mathbf{f} \tag{1.3}$$

The term $\mathbf{p}/\tau$ represents the viscous friction resulting from collisions. In the transport of charges due to an electric field, we have $\mathbf{f} = -e\mathbf{E}$. In the

[3] Georg Ohm (1789-1854).

steady state ($d\mathbf{p}/dt = 0$), $\mathbf{p}(t) = -\tau e\mathbf{E}$, and given eq. 1.1, we recover the local form of Ohm law :

$$\mathbf{J} = \frac{n_e \tau e^2}{m_e}\mathbf{E} = \sigma \mathbf{E} \tag{1.4}$$

where the conductivity σ is proportional to the average time between collisions:

$$\sigma = \frac{n_e \tau e^2}{m_e} \tag{1.5}$$

For metallic elements, conductivity values at room temperature vary between 1 and 20 $\mu\Omega$·cm. Given that typical conduction electron concentrations are of the order of $10^{22} - 10^{23}$ cm^{-3}, this gives an estimate for τ of the order of 1 to 10 fs. From the average kinetic energy of an electron in the electron gas model $\left(\frac{1}{2}m_e < v >^2 = \frac{3}{2}k_B T\right)$, an average free path $< v\tau >$ of the order of 0.1 to 1 nm is estimated, which reasonably coincides with typical interatomic distances. This was taken as a validation of the model of Drude. However, as we shall see below, this estimate of the average free length, based on classical statistical mechanics, is not correct.

1.2.1.1 Shortcomings of the Drude's model

As a classical model, Drude's model has notable discrepancies with experimental results, especially at low temperatures. We briefly indicate some of the main problems:

- The model predicts that the heat capacity of the classical electron gas is independent of temperature T, corresponding to the result obtained from the equipartition theorem. According to this theorem, each degree of freedom corresponds to an average energy $k_B T/2$ (k_B being the constant of Boltzmann[4]), so each electron has an average energy $3\,k_B T/2$ and the heat capacity C of the electron gas results $C = 3\,k_B\,n/2$, n being the concentration of electrons. Experimentally, the heat capacity of the electron gas is proportional to the temperature T.

- In the context of the classical kinetic theory underlying Drude's model, the thermal conductivity κ is expected to be proportional to the heat capacity C: $\kappa = k_B\,T(\tau/m)\,C$, where τ is the relaxation time. Given eq. 1.5, it follows that the relation between thermal conductivity and electrical conductivity (known as the *Wiedemann-Franz ratio*[5]) is expected to be proportional to temperature (Wiedemann-Franz law), the slope being a constant independent of the material, known as the *Lorenz*[6] *number*, L:

[4]Ludwig Boltzmann (1844-1906).
[5]Gustav Wiedemann (1826-1899); Rudolph Franz (1826-1902).
[6]Ludvig Lorenz (1829-1891).

$L = \kappa/(\sigma T) = 3 k_B^2/(2 e^2) \sim 1.1 \times 10^{-8}$ $W\Omega K^{-2}$. Experimentally, the Wiedemann-Franz law fails below room temperature. Moreover, as we saw above, it is based on an incorrect assumption for the heat capacity.

- Drude's model is also unable to explain the experimental observations of Hall[7] effect measurements, where both positive and negative values are obtained for the sign of the electric charges responsible for conductivity in certain materials.

1.2.2 Quantum model of the free electron gas

As we have just seen, Drude's model relies on the classical kinetic theory of gases, based on classical statistical mechanics. The principles of the quantum statistical mechanics of electrons rely on the fundamental principle of the antisymmetry of the multielectron wavefunction with respect to the exchange of any two electrons. This leads to the principle of exclusion of Pauli[8] and to the corresponding quantum statistical distribution function (the Fermi-Dirac[9] function that we shall review in the next section). The corresponding gas of free electrons is usually designated Fermi gas. It was Sommerfeld that first applied this model to the gas of conduction electrons in solids, together with the relaxation time concept, leading to what is usually known as Sommerfeld model of free electrons in solids. This model reasonably explains metallic behavior. In Sommerfeld's model, each electron is treated independently as a free particle and therefore obeys the corresponding Schrödinger[10] equation:

$$-\frac{\hbar^2}{2m_e}\nabla^2\Psi(x,y,z) = E\Psi(x,y,z) \tag{1.6}$$

As known, this equation can be solved using the method of separation of variables:

$$\Psi(x,y,z) = X(x)Y(y)Z(z) \tag{1.7}$$

The separation of variables leads to three differential equations for the variables x, y and z:

$$-\frac{\hbar^2}{2m_e}\frac{d^2X(x)}{dx^2} = E_xX(x)$$

$$-\frac{\hbar^2}{2m_e}\frac{d^2Y(y)}{dy^2} = E_yY(y)$$

$$-\frac{\hbar^2}{2m_e}\frac{d^2Z(z)}{dz^2} = E_zZ(z) \tag{1.8}$$

[7] Edwin Hall (1855-1938).
[8] Wolfgang Pauli (1900-1958).
[9] Enrico Fermi (1901-1954); Paul Dirac (1902-1984).
[10] Erwin Schrödinger (1887-1961).

where $E = E_x + E_y + E_z$. Each of these differential equations has solutions as plane-wave traveling in both senses:

$$\begin{aligned} X(x) &= X_0 \exp(ik_x x),\ k_x = \pm\sqrt{2m_e E_x}/\hbar \\ Y(y) &= Y_0 \exp(ik_y y),\ k_y = \pm\sqrt{2m_e E_y}/\hbar \\ Z(z) &= Z_0 \exp(ik_z z),\ k_z = \pm\sqrt{2m_e E_z}/\hbar \end{aligned} \tag{1.9}$$

We therefore obtain solutions with the general form of a plane wave:

$$\begin{aligned} \Psi(x,y,z) &= A \exp(i\mathbf{k}\cdot\mathbf{r}) \\ E(\mathbf{k}) &= \frac{\hbar^2 k^2}{2m_e} \end{aligned} \tag{1.10}$$

The normalization constant A is obtained from the condition $\int_0^{L_x}\int_0^{L_y}\int_0^{L_z} |\Psi|^2 dx\,dy\,dz = 1$, resulting in $A = 1/\sqrt{L_x L_y L_z} = 1/\sqrt{V}$, where $V = L_x L_y L_z$ is the volume of the solid (parallelepiped of dimensions L_x, L_y and L_z). k_x, k_y and k_z are the components of the wave number vector, $\mathbf{k}$, proportional to the linear momentum $\mathbf{p}$ of the electron, according to the de Broglie[11] relation:

$$\mathbf{p} = \hbar\mathbf{k} \tag{1.11}$$

where

$$\mathbf{p}\Psi = -i\hbar\nabla\Psi = \hbar\mathbf{k}\Psi \tag{1.12}$$

In this ultra-simplified model, we are describing the behavior of electrons confined within the massive interior of the crystalline solid (usually designated *bulk*), far from the surfaces. The confinement assumes a quantum well model with infinite walls. In principle, one could impose fixed conditions (such as $X(0) = X(L_x) = 0$, L_x being the length of the crystal in the x direction). This would lead to standing-wave solutions, such as those found in guitar strings. In this context, however, the relaxation time model implies that the most adequate description is obtained from progressive waves and not as standing waves, so the use of periodic boundary conditions (also known as Born-von Karman[12] boundary conditions) is found more useful, since it leads to progressive-wave solutions:

$$\begin{aligned} X(x+L_x) &= X(x) \\ Y(y+L_y) &= Y(y) \\ Z(z+L_z) &= Z(z) \end{aligned} \tag{1.13}$$

where $L_x = N_x\, a$, $L_y = N_y\, b$ and $L_z = N_z\, c$ are the crystal dimensions in each of the crystallographic directions; a, b and c are the crystallographic

[11] Louis de Broglie (1892-1987).

[12] Max Born (1882-1970); Theodore von Kármán (1881-1963).

parameters. The boundary conditions lead to the following constraints for k_x, k_y and k_z:

$$k_x = n_x \frac{2\pi}{L_x}, n_x = 0, \pm 1, \pm 2, ...$$

$$k_y = n_y \frac{2\pi}{L_y}, n_y = 0, \pm 1, \pm 2, ...$$

$$k_z = n_z \frac{2\pi}{L_z}, n_z = 0, \pm 1, \pm 2, ... \tag{1.14}$$

The energies accessible to the electrons through eq. 1.10 are limited by the conditions eq. 1.14. Each point (k_x, k_y, k_z) corresponds to a possible energy state. The possible values (k_x, k_y, k_z) thus define a three-dimensional lattice of points. The volume V_{k1} occupied by each point in the respective three-dimensional space («k−space») is:

$$V_{k1} = \frac{(2\pi)^3}{L_x L_y L_z} = \frac{(2\pi)^3}{V} \tag{1.15}$$

where V is, as before, the total volume of the crystal. The lowest energy state of the electron gas corresponds to the situation where the independent particle states with lowest energy are occupied at $T = 0$ K, while respecting Pauli exclusion principle. If there are N (free) electrons in the crystal, this corresponds, in k−space, to the occupation of a volume

$$V_{kN} = \frac{1}{2}(2\pi)^3 \frac{N}{V} = 4\pi^3 n \tag{1.16}$$

where $n = N/V$ is the density of free electrons and we have taken into account that each state (k_x, k_y, k_z) allows two electrons with opposite spins. This volume in k−space at $T = 0$ K can be approximately described by a sphere centered at $\mathbf{k} = 0$ (*Fermi sphere*), and with radius k_F (*Fermi radius*):

$$V_{kN} = \frac{4}{3}\pi k_F^3 \rightarrow k_F = (3\pi^2 n)^{1/3} \tag{1.17}$$

The corresponding *Fermi energy* E_F, *Fermi temperature* T_F and *Fermi velocity* v_F can also be defined as:

$$E_F = E(k_F) = \frac{\hbar^2 k_F^2}{2m_e} \tag{1.18}$$

$$T_F = \frac{E_F}{k_B} \tag{1.19}$$

$$v_F = \frac{\hbar}{m} k_F \tag{1.20}$$

Taking into account that the typical concentrations of conduction electrons are of the order of $10^{22} - 10^{23}$ cm^{-3} = $10^{28} - 10^{29}$ m^{-3}, this allows for an estimate for the corresponding orders of magnitude of k_F, E_F, T_F and v_F:

$$\begin{aligned} k_F &\sim 7 \times 10^9 - 1.4 \times 10^{10}\,\mathrm{m}^{-1} \\ E_F &\sim 2 - 8\,\mathrm{eV} \\ T_F &\sim 0.2 \times 10^5 - 1 \times 10^5\,\mathrm{K} \\ v_F &\sim 8 \times 10^5 - 1.6 \times 10^6\,\mathrm{m\,s}^{-1} \end{aligned} \tag{1.21}$$

These numbers are extraordinarily high compared to those obtained using the equipartition theorem. The Pauli principle leads to an "expansion" of the occupied states up to very high values of energy up to E_F, almost two orders of magnitude higher than the thermal energy $k_B T$ (around 25 meV at room temperature). The corresponding values for the Fermi temperature and Fermi velocity are correspondingly higher.

We can now revisit our estimate of the relaxation time obtained within the Drude model. As we shall discuss more thoroughly in Chapter 3, the relaxation time model within the Fermi gas approximation now implies that the whole Fermi sphere moves across the "k−space" through the action of an applied electric field, with the relaxation mechanism affecting essentially the electrons at the border of the sphere (with $E \sim E_F$). However, the corresponding expression for the conductivity (eq. 1.5) is the same, as well as the corresponding relaxation time of 1–10 fs in metallic systems.

However, as we have just seen, the velocity of the electrons at the border of the Fermi sphere is now much higher than that obtained through the classical equipartition result, leading to a correspondingly higher value of the average free length $< v_F \tau >$ of the electrons before colliding. This can now span up to hundreds of angstroms for the lowest conductivity cases. The question of why conduction electrons manage to progress such a long distance in the solid before colliding is only answered through Bloch theorem, when analyzing the quantum mechanical behavior in a periodic potential.

1.2.2.1 Density of states in 3D

In the quantum statistical model of the Fermi gas of conduction electrons, the electrons occupying the lowest energy independent particle states are essentially «inert», as we shall see in the next section. The electronic properties of the material are essentially conditioned by electrons that occupy numerous states at higher energies, with $E \sim E_F$. The degeneracy of the energy levels plays an important role in the correct statistical description of this system. It is therefore essential to know how the number of states varies with energy. Since the relations 1.14 define a set of states with very close energies, which are typically occupied by a number of electrons of the order of magnitude of Avogadro's number, it is convenient to describe the accessible states in a quasi-continuous approximation.

In this context, the density of states function $g(E)$, which describes the number of states per unit volume and unit energy around the energy E, plays a key role. In order to evaluate the density of states function $g(E)$, we start by considering, in the k–space, the volume $4\pi k^2 dk$, bounded by the spheres of radius k and $k+dk$. The number of states $N(k)dk$ contained in this volume is (attending to eq. 1.15),

$$N(k)dk = \frac{4\pi k^2 dk}{V_{k1}} = \frac{k^2 dk}{2\pi^2} V \tag{1.22}$$

It immediately follows that the number of states per unit volume is:

$$g(k)dk = \frac{k^2}{2\pi^2} dk \tag{1.23}$$

where $g(k)$ is already the density of states function that describes the number of states per unit volume and unit of k around the value k. The transformation into the corresponding density of states function $g_s(E)$ is made taking into account that $g(k)dk = g_s(E)dE$, together with eq. 1.10:

$$g_s(E)dE = \frac{1}{2}\frac{\sqrt{2}}{\pi^2}\frac{m_e^{3/2}}{\hbar^3}\sqrt{E}dE \tag{1.24}$$

Each state can accommodate two electrons with opposite spins, so the number of electrons per unit volume and per unit energy around energy E is twice the number calculated in eq. 1.24:

$$g_e(E)dE = \frac{\sqrt{2}}{\pi^2}\frac{m_e^{3/2}}{\hbar^3}\sqrt{E}dE \tag{1.25}$$

1.2.2.2 2D and 1D density of states

Using the procedure shown above, it is possible to obtain the density of states function for cases where the movement of electrons is confined to a plane or to a line. These are highly relevant situations, as we will see in chapter 6. The corresponding density of states functions are (problem 3):

$$\begin{aligned} \text{in 2D} : g(E)dE &= \frac{m_e}{\pi\hbar^2} dE \\ \text{in 1D} : g(E)dE &= \frac{m_e^{1/2}}{\pi\hbar}\frac{1}{\sqrt{E}} dE \end{aligned} \tag{1.26}$$

1.2.2.3 Fermi-Dirac distribution and the meaning of the chemical potential

Once the density of states has been obtained, it is necessary to consider the probability of occupation of a state of energy E by an electron. This probability is governed by the Fermi-Dirac distribution function, which we now briefly review.

In order to find the distribution function of the particles among the possible states, we have to take into account that an isolated system in thermodynamic equilibrium is characterized by a minimum (Helmholtz[13]) free energy.

$$F = U - TS \rightarrow \text{minimum} \tag{1.27}$$

where T is the temperature, U is the total internal energy and S is the entropy:

$$\begin{aligned} U &= \Sigma_i \epsilon_i \\ S &= k_B \ln W \end{aligned} \tag{1.28}$$

$k_B = 1.380\,649 \times 10^{-23}$ J/K $= 8.617\,333\,262... \times 10^{-5}$ eV/K is the Boltzmann constant, ϵ_i is the energy of the particle i and W is the number of microstates accessible to the system (i.e. the number of combinations of N particles, each at a given energy level ϵ_i, forming a system with a given total internal energy U).

Equation 1.27 expresses the known fact that, at a certain temperature, the system is not limited to the minimum internal energy configuration but, due to thermal excitation, has a space of accessible configurations that grows proportionally to temperature and entropy (which in turn reflects the number of accessible microstates).

Quantum particles (such as electrons in solids) are indistinguishable, which implies that the square modulus of the wave function of N particles is invariant to the exchange of coordinates of any two particles:

$$|\Psi(...\mathbf{r_a}...\mathbf{r_b}...)|^2 = |\Psi(...\mathbf{r_b}...\mathbf{r_a}...)|^2 \rightarrow \Psi(...\mathbf{r_a}...\mathbf{r_b}...) = \pm\Psi(...\mathbf{r_b}...\mathbf{r_a}...) \tag{1.29}$$

The wave function of N quantum particles is thus symmetric for the exchange of coordinates of any two particles:

$$\Psi(...\mathbf{r_a}...\mathbf{r_b}...) = +\Psi(...\mathbf{r_b}...\mathbf{r_a}...) \tag{1.30}$$

or antisymmetrical

$$\Psi(...\mathbf{r_a}...\mathbf{r_b}...) = -\Psi(...\mathbf{r_b}...\mathbf{r_a}...) \tag{1.31}$$

A fundamental result (theorem spin-statistics) establishes the relation between the symmetry of the wave function and the spin: particles of integer spin correspond to symmetrical wave functions and particles of half-integer spin (such as electrons) correspond to antisymmetrical wave functions. These two cases correspond to very different probabilities of state occupation. First of all, a fundamental difference immediately emerges for the case where two particles occupy the *same* state: this is possible for particles of integer spin, but in the case of particles of half-integer spin (such as electrons) this means:

$$\Psi(...\mathbf{r_a}...\mathbf{r_a}...) = -\Psi(...\mathbf{r_a}...\mathbf{r_a}...) \rightarrow \Psi(...\mathbf{r_a}...\mathbf{r_a}...) = 0 \tag{1.32}$$

[13]Hermann von Helmholtz (1821-1894).

This result reflects the well-known *Pauli exclusion principle*, which forbids more than one electron from occupying the same state at the same time. However, it is possible to have several electrons in different states with the same energy (degenerate states).

The probability of occupation of a given state thus depends on the spin of the particle, corresponding to two different distribution functions: the Bose-Einstein[14] function for particles with integer spin (*bosons*) and the Fermi-Dirac function for particles with half-integer spin (*fermions*). The Fermi-Dirac function can be obtained by considering an isolated system with N electrons, where N_1 electrons occupy ϵ_1 energy states with a g_1 degeneracy factor, N_2 electrons occupy ϵ_2 energy states with a g_2 degeneracy factor, etc.

Note that if $g_1 = 1$ (nondegenerate state), then the Pauli exclusion principle states that N_1 cannot be larger than 1. For a given value of g_1 larger than 1, N_1 cannot be larger than g_1. The total number of possibilities for distributing N_1 fermions over g_1 states is thus given by the known expression for combinations:

$$P_1 = \frac{g_1!}{N_1!\,(g_1 - N_1)!} \tag{1.33}$$

The number W of microstates accessible to the system becomes:

$$W = \Pi_i P_i = \Pi_i \frac{g_i!}{N_i!\,(g_i - N_i)!} \tag{1.34}$$

Given eq. 1.27 (where $U = \Sigma_i N_i \epsilon_i$), the free energy of the system results:

$$F = U - TS = \Sigma_i N_i \epsilon_i - k_B T \ln \Pi_i \frac{g_i!}{N_i!\,(g_i - N_i)!}, \tag{1.35}$$

which can be easily rewritten, given Stirling's[15] approximation, as:

$$F = \Sigma_i N_i \epsilon_i - k_B T \Sigma_i \left[\ln g_i! - N_i \ln N_i + N_i - (g_i - N_i)\ln(g_i - N_i) + (g_i - N_i)\right] \tag{1.36}$$

The distribution function is $N_i(\epsilon_i)$ and corresponds to the function $N_i(\epsilon_i)$ that minimizes the free energy F. The aim is to obtain the minimum of the function F (which corresponds to the condition $\delta F = 0$) for a given set of occupation numbers N_i, where $\Sigma_i N_i = N$ and N is constant (so $\delta N = 0$). These two conditions ($\delta F = 0$ and $\delta N = 0$) lead to the equivalent condition:

$$\delta F - \mu \delta N = 0 \tag{1.37}$$

where μ is a parameter to be determined *a posteriori* from condition $\Sigma_i N_i = N$. The physical interpretation of the parameter μ is immediate from eq. 1.37: μ corresponds to the variation in free energy associated with a variation in the number of particles in the system; thus it is the (average) free energy

---

[14] Satyendra Bose (1894-1974); Albert Einstein (1879-1955).

[15] James Stirling (1692-1770). Stirling's approximation is $\ln(N!) \sim N\ln(N) - N$.

associated with each particle. This parameter μ is usually called *chemical potential.*

From eqs. 1.36 and 1.37, it is easy to conclude:

$$\Sigma_i \left[\epsilon_i - k_B T \ln\left(\frac{g_i - N_i}{N_i}\right) - \mu\right] \delta N_i = 0, \tag{1.38}$$

and the occupation N_i follows immediately:

$$N_i = N(\epsilon_i) = \frac{g_i}{1 + \exp\left(\frac{\epsilon_i - \mu}{k_B T}\right)} \tag{1.39}$$

The Fermi-Dirac distribution is thus defined as:

$$f(\epsilon_i) = \frac{N(\epsilon_i)}{g_i} = \frac{1}{1 + \exp\left(\frac{\epsilon_i - \mu}{k_B T}\right)} \tag{1.40}$$

Figure 1.1 illustrates the behavior of the Fermi-Dirac distribution for a given value of the chemical potential (here taken as $\mu = 1$ in arbitrary units) and for some representative values of the thermal energy $k_B T$. For very small values of thermal energy $k_B T$ ($k_B T = 0.01$ in the example of figure 1.1), all energy levels E below the value of the chemical potential are occupied ($f(E) = 1$) and the energy levels above the value of the chemical potential are all empty. This corresponds to the limit of the Fermi-Dirac distribution for $T \to 0$: $f(E) = 1$ for $E < \mu$ and $f(E) = 0$ for $E > \mu$. At very low temperatures, the chemical potential thus represents the maximum energy value of the occupied levels (also called *Fermi energy*). In the case of the Sommerfeld model for metallic behavior that we have seen in section 1.2.2, the chemical potential simply corresponds to the Fermi energy at $T = 0$ K. Note, however, that the chemical potential is a thermodynamic quantity, of a

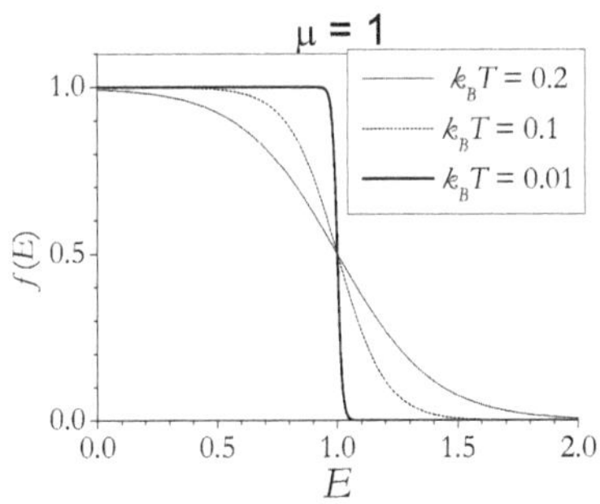

FIGURE 1.1
Representation of the Fermi-Dirac distribution for a given value of the chemical potential (here taken as $\mu = 1$ in arbitrary units) and for some representative values of the thermal energy $k_B T$.

statistical nature, so this maximum value expressed in the chemical potential *does not necessarily correspond* to an energy level: this detail is of the utmost importance in semiconductor physics.

Another very important aspect expressed in figure 1.1 is the progressive occupation of energy levels above the chemical potential as the thermal energy k_BT increases. The curve with $k_BT = 0.1$ clearly expresses how this happens: electrons occupying the energy levels just below the chemical potential μ are promoted to levels the unoccupied levels just above μ. However, the probability of occupation decreases rapidly, with $f(E) = 0.5$ for $E = \mu$; for $E > \mu$, the classical Boltzmann distribution is recovered (for $\exp(E - \mu)/k_BT >> 1$):

$$f(E) \sim \exp\left(-\frac{E - \mu}{k_BT}\right) \tag{1.41}$$

The thermal excitation process thus involves only a fraction of all the electrons, roughly corresponding to (ignoring degeneracy effects):

$$\frac{\int_\mu^\infty f(E)dE}{\int_0^\infty f(E)dE} \tag{1.42}$$

This fraction is in general very small, since $\mu >> k_BT$ (as we have seen above, the Pauli principle forces the occupation of energy levels much above the ground state) and can therefore be easily estimated, considering the result 1.41 for $E > \mu$ and $f(E) \sim 1$ for $E < \mu$:

$$\frac{\int_\mu^\infty f(E)dE}{\int_0^\infty f(E)dE} \sim \frac{k_BT}{\mu} \tag{1.43}$$

The fraction of thermally excited electrons at a given temperature T is thus proportional to the temperature, in contrast to the classical result. It also corresponds to a very small value, taking into account that μ is of the order of the Fermi energy ($\sim$ eV), whereas $k_BT \sim 10^1$ meV. This implies a value of the heat capacity of the electronic gas which is proportional to temperature and is much reduced with respect to the classical result.

1.2.3 Electrons in periodic potentials: theorem of Bloch

The statistical model of the quantum gas of free electrons (Sommerfeld model) introduces a qualitative change in the approach to the behavior of electrons in solids and implies that only electrons in a narrow range of energies around the Fermi energy (approximately k_BT wide) play a relevant role in the physical properties of materials. This extremely simplified model is just a starting point for the quantum treatment of electronic properties.

The next step is, still in the context of a model of independent electrons, to introduce the periodic nature of the electrostatic potential, which reflects the periodicity in the spatial arrangement of atoms in crystal lattices.

The corresponding time-independent Schrödinger equation is now:

$$\left[-\frac{\hbar^2}{2m_e}\nabla^2 + U(\mathbf{r})\right]\Psi(x,y,z) = E\Psi(x,y,z) \tag{1.44}$$

where the potential $U(\mathbf{r})$ has the translation symmetry associated with the crystal lattice:

$$U(\mathbf{r}) = U(\mathbf{r}+\mathbf{R}) \tag{1.45}$$

and $\mathbf{R} = n_1\mathbf{a_1} + n_2\mathbf{a_2} + n_3\mathbf{a_3}$ is a crystal lattice vector with primitive vectors $\mathbf{a_1}$, $\mathbf{a_2}$ and $\mathbf{a_3}$. Recall that in the case of $U = 0$, the wavefunction has the typical plane-wave shape (eqs. 1.10 and 1.11):

$$\begin{aligned} \Psi(x,y,z) &= \frac{1}{\sqrt{V}}\exp(i\mathbf{k}\cdot\mathbf{r}) \\ E(\mathbf{k}) &= \frac{\hbar^2 k^2}{2m_e} \\ \mathbf{p} &= \hbar\mathbf{k} \end{aligned} \tag{1.46}$$

In the case of the periodic potential 1.45, the electron density (proportional to the square modulus of the wave function Ψ) will necessarily reflect the translational symmetry of the crystal lattice:

$$|\Psi(\mathbf{r})|^2 = |\Psi(\mathbf{r}+\mathbf{R})|^2 \tag{1.47}$$

It can be shown that, in these circumstances, the wave function is simply the product of a plane wave $\exp(i\mathbf{k}\cdot\mathbf{r})$ by a function $u_{\mathbf{k}}(\mathbf{r})$ with the translation symmetry of the crystal lattice:

$$\begin{aligned} \Psi_{\mathbf{k}}(\mathbf{r}) &= u_{\mathbf{k}}(\mathbf{r})\exp(i\mathbf{k}\cdot\mathbf{r}) \\ u_{\mathbf{k}}(\mathbf{r}) &= u_{\mathbf{k}}(\mathbf{r}+\mathbf{R}) \end{aligned} \tag{1.48}$$

This result is known as *Bloch theorem.*[16]

1.2.3.1 Meaning of k in Bloch theorem; semiclassical model

In the Sommerfeld model, the electron wave function corresponds, as we mentioned, to plane waves of the form $\exp(i\mathbf{k}\cdot\mathbf{r})$, where $\mathbf{p} = \hbar\mathbf{k}$ represents the linear momentum of the electron. However, the wave function expressed in the form 1.48 is not an eigenfunction of linear momentum (problem 5) and the parameter $\mathbf{k}$ that characterizes it does not obey condition 1.11.

The wave function expressed in the form 1.48 is, however, an eigenfunction of the Hamiltonian.[17] This means, according to basic quantum mechanics, that

[16] Felix Bloch (1905-1983). The proof of Bloch theorem can be found in any textbook on Quantum Mechanics or Solid State Physics. See for example Jasprit Singh, *Quantum Mechanics: Fundamentals & Applications to Technology*, pp. 100-105.

[17] William Hamilton (1805-1865).

the eigenstates of the linear momentum are not eigenstates of the Hamiltonian and are therefore not stationary states, thus constituting a different basis. It is possible, however, to write one basis as a function of the other and to write the well-defined energy states as a function of the well-defined linear momentum states (and vice versa).

This phenomenon is well known from wave theory: wave packets are generally described by the superposition of waves with different frequencies $\nu = \omega/2\pi$ (and energy E) and wavelengths $\lambda = 2\pi/k$ (and linear momentum $\mathbf{p}$). In this case, the propagation speed of the wave packet is the well-known group velocity:

$$v_g = \frac{d\omega}{dk} \tag{1.49}$$

or, in 3 dimensions:

$$v_g = \nabla_{\mathbf{k}}\, \omega(\mathbf{k}) \tag{1.50}$$

It can be shown[18] that an analogous expression is valid for Bloch electrons described by the wave function 1.48 and characterized by the parameter $\mathbf{k}$ and the energy $E(\mathbf{k}) = \hbar\omega$. The group velocity of these electrons can be written in the form

$$\mathbf{v_g} = \frac{1}{\hbar}\nabla_{\mathbf{k}} E(\mathbf{k}) \tag{1.51}$$

In addition to this expression for the velocity of Bloch electron, it is also possible to write an equation analogous to Newton's[19] second law. In this semi-classical approach, the wavelength of the applied field is much larger than the spread of the electron's wavepacket, which by its turn is much larger than the lattice constant.[20] We consider that the change in energy dE of an electron subject to the external force $\mathbf{F_{ext}}$ during the time interval dt is:

$$dE = \mathbf{F_{ext}} \cdot \mathbf{v}dt \tag{1.52}$$

This change in energy dE can also be written taking eq. 1.51 into account:

$$dE = \nabla_{\mathbf{k}} E(\mathbf{k}) \cdot \mathbf{dk} = \hbar\mathbf{v} \cdot \mathbf{dk} \tag{1.53}$$

From these two equations, it immediately follows that $\hbar\mathbf{v} \cdot \mathbf{dk} = \mathbf{F_{ext}} \cdot \mathbf{v}dt$, so that

$$\hbar\frac{d\mathbf{k}}{dt} = \mathbf{F_{ext}} \tag{1.54}$$

The expression 1.54 is remarkable because it indicates that the time evolution of the parameter $\mathbf{k}$ is determined exclusively by the external force applied to the Bloch electron, and that the effects of the crystal potential are not explicitly involved. The formal similarity of eq. 1.54 to Newton's second law justifies that $\mathbf{k}$ is known as *pseudo-momentum* or *crystalline momentum*.

---

[18] Neil D. Ashcroft and D. David Mermin, *Solid State Physics*, Appendix E.
[19] Isaac Newton (1643-1727).
[20] Neil D. Ashcroft and D. David Mermin, *Solid State Physics*, p. 217.

How, then, does the crystal potential affect the electron dynamics in this model? Consider the rate of change in velocity over time (the acceleration):

$$\frac{d\mathbf{v}}{dt} = \frac{1}{\hbar}\frac{d}{dt}\nabla_{\mathbf{k}}E(\mathbf{k}) \tag{1.55}$$

Writing it, for simplicity, in one dimension:

$$\frac{dv}{dt} = \frac{1}{\hbar}\frac{d}{dt}\frac{dE}{dk} = \frac{1}{\hbar}\frac{d^2E}{dk^2}\frac{dk}{dt} \tag{1.56}$$

Taking eq. 1.54 into account, it results:

$$\frac{dv}{dt} = \frac{1}{\hbar^2}\frac{d^2E}{dk^2}F_{\text{ext}} = \frac{F_{\text{ext}}}{m^*} \tag{1.57}$$

Equation 1.57 represents the analog of Newton's second law written in the form $a = F/m$. Once again, only the external force is involved in this equation, but instead of the mass of the electron, the following expression has to be considered:

$$\frac{1}{m^*} = \frac{1}{\hbar^2}\frac{d^2E(\mathbf{k})}{dk^2} \tag{1.58}$$

m^* is therefore designated *effective mass* and is a quantity which includes the effects of the periodical crystalline potential in the dynamics of the Bloch electron. When generalizing to three dimensions, the effective mass tensor is defined as:

$$\frac{1}{m^*_{ij}} = \frac{1}{\hbar^2}\frac{\partial^2E(\mathbf{k})}{\partial k_i \partial k_j} \tag{1.59}$$

1.2.4 Periodic crystal lattices: basic concepts

In order to move forward with the understanding of the basic aspects of the electronic structure of crystalline solids, we need to review some concepts associated with the periodicity of crystalline lattices, such as the concepts of reciprocal lattice, Wigner-Seitz[21] cell of the reciprocal lattice and Brillouin[22] zones. These concepts reflect the fact that the periodic potential can be developed as a Fourier[23] series in real space, implying the possibility of defining a reciprocal space of "spatial frequencies": the reciprocal lattice.

Recall that the translational periodicity of the crystal lattice is expressed by the lattice primitive vectors $\mathbf{a_1}$, $\mathbf{a_2}$ and $\mathbf{a_3}$, with any lattice point $\mathbf{R}$ being expressed as

$$\mathbf{R} = n_1\mathbf{a_1} + n_2\mathbf{a_2} + n_3\mathbf{a_3} \tag{1.60}$$

where n_1, n_2 and n_3 are integers.

[21] Eugene Wigner (1902-1995); Frederick Seitz (1911-2008).
[22] Léon Brillouin (1889-1969).
[23] Jean-Baptiste Fourier (1768-1830).

The potential to which the electrons are subjected has the spatial periodicity of the crystal lattice,

$$V(\mathbf{r}+\mathbf{R}) = V(\mathbf{r}), \tag{1.61}$$

and can therefore be written in the form of a Fourier series in the reciprocal lattice:

$$V(\mathbf{r}) = \sum_{\mathbf{G}} V_{\mathbf{G}} \exp\left(i\mathbf{G}\cdot\mathbf{r}\right), \tag{1.62}$$

where $\mathbf{G}$ is a vector of the reciprocal lattice. Equations 1.61 and 1.62 imply:

$$\exp\left(i\mathbf{G}\cdot\mathbf{R}\right) = 1 \rightarrow \mathbf{G}\cdot\mathbf{R} = 2l\pi \tag{1.63}$$

where l is an integer. The vector $\mathbf{G}$ can be written as a function of the respective (as yet unknown) basis vectors $\mathbf{A_1}$, $\mathbf{A_2}$ and $\mathbf{A_3}$:

$$\mathbf{G} = m_1\mathbf{A_1} + m_2\mathbf{A_2} + m_3\mathbf{A_3} \tag{1.64}$$

where m_1, m_2 and m_3 are integer coefficients to be determined.

Since $\mathbf{R} = \sum_j n_j\,\mathbf{a_j}$, the relation 1.63 implies that

$$\sum_{i,j} \left(m_i\,\mathbf{A_i}\cdot n_j\,\mathbf{a_j}\right) = 2l\pi \tag{1.65}$$

We hence obtain the usual definition of the primitive vectors of the reciprocal lattice, from the corresponding primitive vectors of the real lattice:

$$\mathbf{A_i}\cdot\mathbf{a_j} = 2\pi\delta_{ij} \tag{1.66}$$

The Wigner-Seitz unit cell of the reciprocal lattice is usually called *first Brillouin zone* and has a volume V_k, in the reciprocal space, given by:

$$V_k = \mathbf{A_1}\cdot\left(\mathbf{A_2}\times\mathbf{A_3}\right) \tag{1.67}$$

1.2.4.1 Number of states contained in the first Brillouin zone

One of the most important properties of the first Brillouin zone is the number of states accessible to the electrons contained therein, which, as we will show below, corresponds to the number of unit cells in the crystalline solid. To do this, consider a wave function written in plane wave form (one of the possible bases, so no generality is lost):

$$\phi(\mathbf{r}) = A\exp\left(i\mathbf{k}\cdot\mathbf{r}\right) \tag{1.68}$$

where

$$\mathbf{k} = \sum_i l_i\mathbf{A_i} \tag{1.69}$$

is a vector of the reciprocal lattice. Using the periodic boundary conditions 1.13, we have

$$\phi\left(\mathbf{r}+\sum_j N_j \mathbf{a_j}\right)=\phi(\mathbf{r}) \tag{1.70}$$

where N_1, N_2 and N_3 are the number of unit cells in the direction of the primitive vectors $\mathbf{a_1}$, $\mathbf{a_2}$ and $\mathbf{a_3}$, respectively. It follows that

$$\exp\left(i\mathbf{k}\cdot\sum_j N_j \mathbf{a_j}\right)=1 \rightarrow \exp\left(i\sum_j N_j \mathbf{k}\cdot\mathbf{a_j}\right)=1 \tag{1.71}$$

and, given the succession of arguments presented in eqs. 1.63–1.66, it follows that the minimal vectors $\mathbf{k}$ result:

$$\mathbf{k}=\frac{1}{N_j}\mathbf{A_j} \tag{1.72}$$

We thus have, in general:

$$\mathbf{k}=\sum_j m_j\frac{1}{N_j}\mathbf{A_j} \tag{1.73}$$

The «volume» V_{state} associated with each state within the first Brillouin zone is:

$$V_{state}=\mathbf{k_1}\cdot(\mathbf{k_2}\times\mathbf{k_3})=\frac{1}{N_1N_2N_3}\mathbf{A_1}\cdot(\mathbf{A_2}\times\mathbf{A_3}) \tag{1.74}$$

Comparison with the eq, 1.67 immediately shows that

$$V_{state}=\frac{1}{N_1N_2N_3}V_k \Longleftrightarrow V_k=N_1N_2N_3V_{state} \tag{1.75}$$

The total volume of the first Brillouin zone is therefore the product of the volume of an individual state by the total number of unit cells in the crystal. There are therefore as many states in the first Brillouin zone as there are unit cells in the crystal.

1.3 Band structure

As we have seen, Bloch theorem allows us to draw far-reaching conclusions about the behavior of electrons in periodic potentials. Solving Schrödinger equation (eq. 1.44) and determining the corresponding E energy levels, however, remains a formidable task. In this section we will look at some simple approximations that reveal some fundamental aspects of the electronic structure of crystalline solids. The appearance of energy intervals accessible to electrons (*energy bands*), separated by gaps of forbidden energies (*bandgap*) is already evident in a one-dimensional model of a periodic potential (Kronig-Penney[24] model, problem 10). A simplified analysis of the Schrödinger equation, which we will outline below, allows us to delve deeper into the physical meaning of the bandgap. Finally, taking into account the number of electronic states contained in each band, we will present the distinction between metals and insulators from an electronic point of view.

1.3.1 Brillouin zone boundary analysis

As we have seen, the spatial periodicity of the crystal lattice means that it is possible to use Fourier analysis techniques as a function of the respective spatial frequencies, resulting in the definition of the reciprocal lattice. Equation 1.62 expresses the development of the Fourier series potential in the reciprocal lattice. In turn, the wave function can be written as a combination of the vectors $\mathbf{k}$:

$$\Psi(\mathbf{r}) = \sum_{\mathbf{k}} C_{\mathbf{k}} \exp\left(i\,\mathbf{k}\cdot\mathbf{r}\right), \tag{1.76}$$

These developments in Fourier series make it possible to rewrite the Schrödinger equation in the coefficients $V_{\mathbf{G}}$ and $C_{\mathbf{k}}$, resulting in:

$$(\lambda_{\mathbf{k}-\mathbf{G}} - E)C_{\mathbf{k}-\mathbf{G}} + \sum_{\mathbf{G}'} V_{\mathbf{G}'-\mathbf{G}} C_{\mathbf{k}-\mathbf{G}'} = 0 \tag{1.77}$$

where $\lambda_{\mathbf{k}} = \hbar^2|\mathbf{k}|^2/2m$. This set of equations in the coefficients allows to obtain an approximate solution on the border of the Brillouin zone ($\mathbf{k} \sim \mathbf{G}/2$). Very schematically,[25] it is possible to limit the analysis on the Brillouin zone boundary to the coefficients $C_{\mathbf{G}/2}$ and $C_{-\mathbf{G}/2}$, resulting in a set of two equations leading to the determinant

$$\begin{vmatrix} \lambda - E & V \\ V & \lambda - E \end{vmatrix} = 0 \Leftrightarrow (\lambda - E)^2 - V^2 = 0 \Leftrightarrow E_{\pm} = \lambda \pm V \tag{1.78}$$

This results in two possible energy values on the border of the Brillouin zone at $\lambda = \hbar^2 G^2/4$, separated by the energy $2V$:

$$E_{\pm} = \frac{\hbar^2}{2m}\frac{G^2}{4} \pm V \tag{1.79}$$

In the vicinity of the Brillouin area, the determinant is written as:

$$\begin{vmatrix} \lambda_{\mathbf{k}} - E & V \\ V & \lambda_{\mathbf{k}-\mathbf{G}} - E \end{vmatrix} = 0, \tag{1.80}$$

[24] Ralph Kronig (1904-1995); William Penney (1909-1991).

[25] A detailed analysis can be found in Jasprit Singh, *Quantum Mechanics: Fundamentals & Applications to Technology*, pp. 288-294.

resulting in the solution

$$E_{\pm}(\mathbf{k}) \sim \lambda \pm V + \frac{\hbar^2 k^2}{2m}\left(1 \pm \frac{2\lambda}{V}\right) \tag{1.81}$$

This results in a parabolic dispersion relation at the boundary of the Brillouin zone, with a positive curvature for the higher energy E_+ and a negative curvature for the lower energy E_-. The curvature can be obtained immediately, resulting in the effective mass (for the case $|V| << 2\lambda$):

$$m^* = m\frac{1}{1 \pm \frac{2\lambda}{V}} \sim \pm\frac{V}{2\lambda}m \tag{1.82}$$

Note that the two energy values $E_\pm$ obtained at the border of the Brillouin zone correspond to $\mathbf{k} \sim \mathbf{G}/2$. For these wavelengths, which correspond to the interatomic distances, the effects of electron reflection/diffraction generate stationary waves, for which there is no propagation. The two different energy values correspond to the two possible stationary waves with lattice symmetry: one where the wave function is maximum at the atomic positions (of energy E_-) and another where it is minimum (of energy E_+).[26]

1.3.2 Band representation and classification of solids

The energy bands correspond to the possible values of the electron energy (solution of the Schrödinger equation) as a function of the possible values of the pseudo-momentum **k**. A complete graphical representation of a band is obviously out of the question (it would require 4 dimensions!), so it is usual to represent the energy values in specific directions in the k−space. Typically, directions of high symmetry, which correspond to the extreme values of the bands and thus allow a global view in a two-dimensional graph, are chosen.

Before presenting real results for bands in solids, we review the case of a one-dimensional chain of atoms, separated by a distance of a, where there is a probability, characterized by the parameter γ, of an electron hopping to a nearby atom. It is possible to show, using the tight-binding model,[27] that the corresponding one-dimensional band has the general form

$$E(k) = E_0 - 2\,\gamma\,\cos(a\,k) \tag{1.83}$$

In figure 1.2 we represent some one-dimensional bands, in the entire k−space. The representation of figure 1.2 (i) highlights the periodic aspect of the bands in the k−space (with the period $2\pi/a$ corresponding to the basis vector of the one-dimensional reciprocal lattice). This type of representation is called *periodic representation* and is clearly redundant.

[26]For a more detailed analysis, see Charles Kittel, *Introduction to Solid State Physics*, 7th ed., Chap. 7.

[27]This model can be found in any introductory book on band structure. See, for example, John Singleton, *Band theory and electronic properties of solids*, Chap. 4.

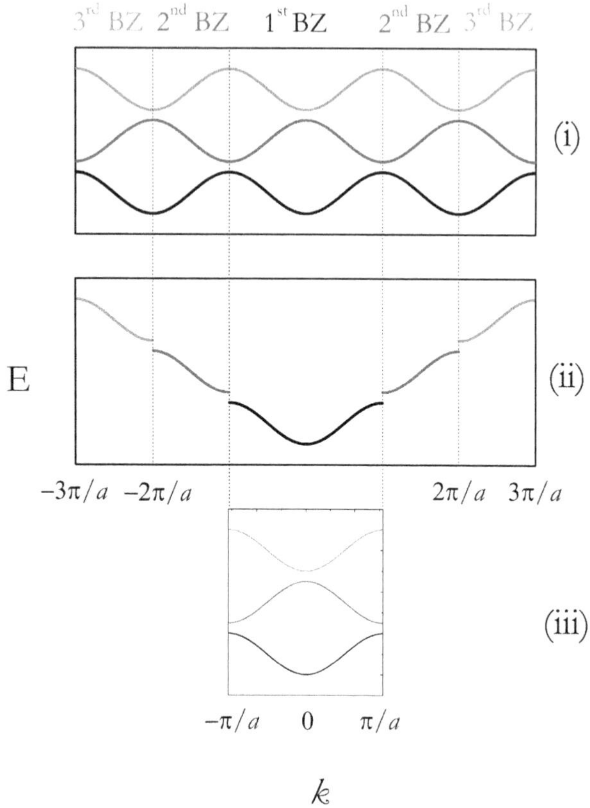

FIGURE 1.2
One-dimensional band representations: (i) periodic; (ii) extended; (iii) reduced.

The periodicity of the bands in the reciprocal lattice implies that all the information is in the first Brillouin zone, as illustrated in figure 1.2 (iii). This type of representation is called *reduced representation* and is widely used due to its compact nature. In the representation of figure 1.2 (iii) the bandgaps between the bands are evident, as well as the parabolic nature of the bands near the energy minima/maxima; in addition, the different curvature of the bands at the border of the Brillouin zones is evident, as is the zero slope of the bands at the border of the Brillouin zone, reflecting the zero velocity of the electrons due to the strong reflection effects that occur there, with the consequent formation of stationary waves.

Sometimes, a third type of representation is used, where the bands with increasing energy are represented in successive Brillouin zones (*extended representation*, figure 1.2 (ii)). In figure 1.2 (ii) it is clear, for example, that the

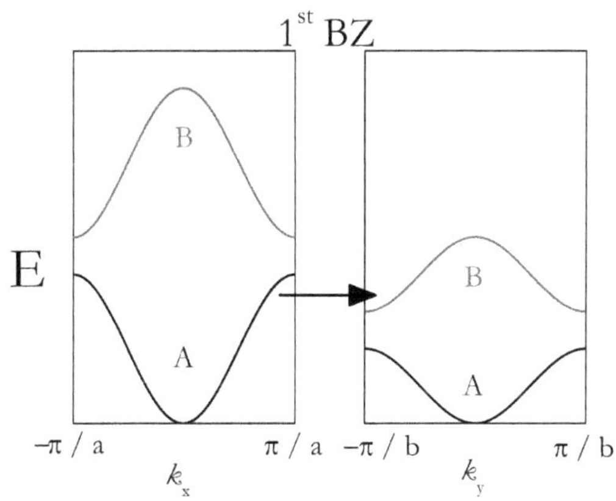

FIGURE 1.3
Schematic representation of the band structure of a divalent metal or a semi-metal.

dispersion curve of the tight-binding model is similar to a parabola (corresponding to the free electron model) where continuity is broken at the borders of the Brillouin zones, with the appearance of bandgaps.

1.3.2.1 Classification of solids

As we have seen, each Brillouin zone (and therefore each band) has as many states as there are unit cells in the crystal, which correspond to twice the number of electrons. Thus, the band corresponding to a crystal made up of monovalent atoms, with one atom per unit cell and therefore one electron per unit cell (for example Cu), is half-filled. The application of a weak constant electric field corresponds to applying a linear potential to the material, which elevates the energy at one end with respect to the other. This allows the electrons to start flowing to the low energy end by occupying the levels that are empty. This gives rise to an electric current and to metallic behavior.

In the case of a crystal made up of divalent atoms, the situation requires a closer look. In the one-dimensional case shown in figure 1.2, there is no doubt: the band is completely filled. By applying a weak electric field, there are no adjacent unoccupied levels in k−space to which the electrons in this band can be promoted and therefore no electric current can be generated: these materials are insulators. The lower fully occupied band is called the *valence band* and the upper unoccupied band is called *conduction band.* In the common three-dimensional case (and even in the two-dimensional case), a situation like the one shown in figure 1.3 may occur, where the bands have different energies in different directions of the k−space.

In the case of figure 1.3, the lower energy states of the B band in the k_y direction have lower energy than the higher energy states of the A band

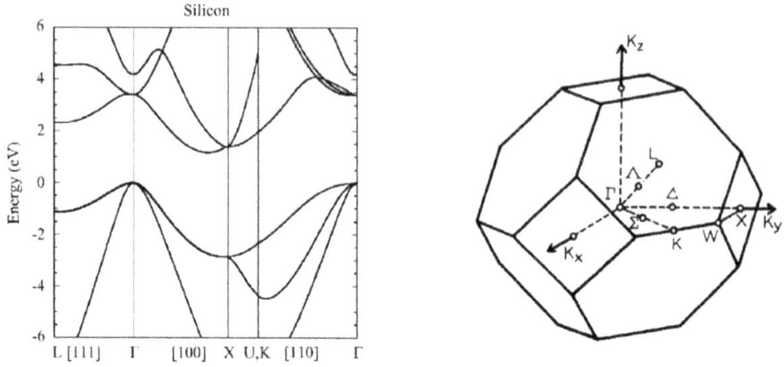

FIGURE 1.4
Left: Band structure of Si calculated at $T = 0$ K. (Figure reproduced with permission from S. Richard, F. Aniel, and G. Fishman, *Energy-band structure of Ge, Si, and GaAs: A thirty-band k · p method*, Phys. Rev. B **70**, 235204 (2004). *Copyright* (2004) of the American Physical Society). **Right:** Brillouin zone of the face-centered lattice, with some positions and lines of high symmetry indicated. (Figure reproduced with permission: P. A. Temple e C. E. Hathaway, *Multiphonon Raman Spectrum of Silicon*, Phys. Rev. B **7**, 3685 (1973). *Copyright* (1973) of the American Physical Society).

in the k_x direction, so the A band is not completely filled in this direction and the B band is partially filled in the k_y direction. This type of material also conducts electricity, but with generally lower conductivity due to the reduced concentration of charge carriers. These materials are therefore also metals (e.g. Zn). However, when the overlap of the bands occurs in such a narrow energy range (of the order of meV) that the concentration of charge carriers is much reduced, the conductivity, in addition to being extremely low, becomes strongly temperature dependent, and these materials are designated *semi-metals* (e.g. Bi).

1.3.2.2 Band structure of silicon

We will now consider the paradigmatic case of the semiconductor *par excellence*: silicon. Its band structure is shown in figure 1.4. As it is a three-dimensional crystal, the bands are represented in certain highly symmetrical directions of the first Brillouin zone, which is also represented on the right side of figure 1.4 along with an indication of the points of high symmetry. In the perfect crystal and at low temperatures, the bands with energy below zero are all occupied and are therefore valence bands. The following bands begin only after a gap of about 1 eV and are conduction bands. Perfect silicon is therefore an insulator: how does it become a semiconductor?

A common argument is to identify semiconductor materials with those insulators where the bandgap is so small that the thermal energy $k_B T$ is sufficient to justify a significant occupation of the lower energy states of the conduction band. However, this argument does not stand up to the fact that at room temperature the thermal energy is only around 25 meV, which is clearly not enough to create a concentration of electrons in the conduction band of pure silicon to significantly affect its conductivity. As we will see in more detail in the following chapter, this type of process becomes important, in silicon, as well as in most semiconductors, at temperatures much higher than room temperature.

The hallmark of semiconductors is therefore not so much the value of the bandgap, but rather the existence of mechanisms capable of promoting electrons to the conduction band (or, as we shall see, holes to the valence band). These mechanisms are based on the presence of impurities and/or defects that introduce spurious levels into the band structure of the pure semiconductor. The physics of semiconductors is thus closely linked to the physics of defects/impurities in semiconductors.[28]

As for the band structure of silicon, the following should also be noted:

- The maximum of the valence band occurs at $\mathbf{k} \equiv (0\,0\,0)$ (Γ point), but the minimum of the conduction band occurs at the X points (of which there are six equivalent points in the Brillouin zone, two on each axis). The bandgap therefore does not correspond to points with the same crystal momentum, and is called an *indirect bandgap.* As we shall see in Chapter 4, this circumstance has important consequences for optical properties.

- The minimum of the conduction band occurs at the six equivalent X points, which implies that the excitation of electrons into this band occurs at these points. However, the curvature of the conduction bands in the directions perpendicular to the axis ($X-U$ or $X-W$ directions) is different from the curvature in the $X-\Gamma$ direction parallel to the axis, giving rise to different effective masses and to an anisotropic and axially symmetric filling.

- The top of the valence band corresponds to two bands with different curvatures, which translate into different effective masses, giving rise to the corresponding designation of *light holes* and *heavy holes.*

[28]This crucial point justifies the initial difficulties in establishing solid foundations for semiconductor physics, leading Pauli to state: *Über Halbleiter sollte man nicht arbeiten, das ist eine Schweinerei, werweiss ob es überhaupt Halbleiter gibt.* ("We shouldn't work on semiconductors, that is a mess, who knows whether any semiconductor exists.") Wolfgang Pauli, Letter to Rudolph Peierls, September 29, 1931, *apud* M. Grundmann, *The Physics of Semiconductors*, 2nd edition, p. 185.

1.3.3 The electronic hole

As mentioned above, at high temperatures, the thermal excitation of electrons from the valence band to the conduction band becomes important in silicon. This process leaves an unoccupied state in the valence band, called *electronic hole* or simply *hole*.[29] As we will see in the next chapter, in semiconductors the presence of impurities and/or defects can also give rise to holes in the valence band by promoting electrons to levels in the bandgap.

These unoccupied states play a very important role, particularly in electrical conduction, because they provide a mechanism for the electrical conduction of valence electrons, albeit in typically low concentrations. Note that valence band electrons are electrons involved in bonds between atoms. The creation of a hole in the valence band thus corresponds to the removal of an electron from one of these bonds, so that locally there is the creation of a *positive* charge: the dynamics of holes thus corresponds to the dynamics of positive charges, which immediately explains the mysterious positive signs observed in the Hall coefficients of various semiconductors. It is much more convenient to describe the dynamics of electronic holes *per se* than to describe the dynamics of the other electrons in the valence band. This implies describing an unoccupied state, whose properties must be related to the physical parameters of the removed electron.

1.3.3.1 a) $\mathbf{k}_h = -\mathbf{k}_e$

Note that the symmetry of the first Brillouin zone implies that $\sum_i \mathbf{k}_{ei} = 0$. By removing the electron with crystal momentum $\mathbf{k}'_e$, we get $\mathbf{k}'_e + \sum'_i \mathbf{k}_{ei} = 0$, where $\sum' \mathbf{k}_{ei}$ designates the remaining crystal momentum in the valence band, which can therefore be defined as the crystal momentum associated with the electron hole: $\mathbf{k}_h = -\sum' \mathbf{k}_{ei} = -\mathbf{k}_e$.

1.3.3.2 b) $E_h(\mathbf{k}_h) = -E_e(\mathbf{k}_e)$

A similar idea applies to the sum of the energies of the electrons in the valence band. This sum is a constant and the value of this constant does not have any particular physical meaning, since only energy differences have physical meaning; this constant can be taken to be zero, resulting in $E_h(\mathbf{k}_h) = -E_e(\mathbf{k}_e)$. A very important aspect of this relation is that the holes with the lowest energy are those corresponding to the top of the valence band, since they correspond to the electrons for which the lowest energy is required for promotion to a level in the bandgap or conduction band. From these relations, the velocity of the hole immediately follows as $\mathbf{v}_h(\mathbf{k}_h) = \mathbf{v}_e(\mathbf{k}_e)$ and its effective mass as $m_h^*(\mathbf{k}_h) = -m_e^*(\mathbf{k}_e)$, as well as the dynamics due to the action of external forces (namely electric and magnetic fields), which we leave as an exercise (problem 8).

[29] This hole should not be confused with a common defect corresponding to the lack of an atom in the lattice: the *atomic vacancy*.

Problems

1. Check eqs. 1.8, 1.9, 1.10 and 1.12.
2. Obtain the density of states 1.24 from the relation 1.23.
3. Obtain the 2D and 1D density of states functions (eq. 1.26).
4. From the Fermi-Dirac distribution (eq. 1.40), show that the fraction of electrons that, at a given temperature T, occupy energy levels above the chemical potential is proportional to the temperature (eq. 1.43).
5. Check that the wave function 1.48 is not a function of linear momentum and that, therefore, $\mathbf{p}\Psi_{\mathbf{k}}(\mathbf{r}) = -i\hbar\nabla\Psi_{\mathbf{k}}(\mathbf{r}) \neq \hbar\mathbf{k}\Psi_{\mathbf{k}}(\mathbf{r})$.
6. The specific mass of copper is 8.95 g.cm^{-3} and its atomic weight is 63.54. The electrical resistivity at temperature is $\rho = 1.51 \times 10^{-6}\,\Omega.cm$.
   (a) Calculate the concentration of conduction electrons.
   (b) Consider that an electric field $E = 100$ V cm^{-1} is applied to a portion of the metal material of section A and length L. Calculate the drift velocity of the electrons and compare it with the average thermal velocity. Calculate the average time between collisions or relaxation time, τ. Use Drude model.
   (c) Using the Sommerfeld model for free electrons, calculate the Fermi velocity, the Fermi energy and the Fermi temperature.
   (d) Calculate the mean free path, l. Compare this value with the one obtained using Drude model.
7. Consider a rectangular lattice of sides $a = 3$ Å and $b = 4$ Å, containing an atom A in position $(0, 0)$ and an atom B in the position $\left(\frac{1}{2}, \frac{1}{2}\right)$. Suppose that the dispersion relation of electrons is adequately described by the free electron model.
   (a) Approximately represent the dispersion curves in the directions [10], [01] and [11].
   (b) Suppose that each AB molecule gives up two electrons to the conduction band. What relation must exist between the width of the forbidden band at the point k_{ZB} $[\Delta U(k_{ZB})]$ and the lattice parameters for the material to be an insulator? In the opposite hypothesis (the material is a conductor), represent the shape of the Fermi surface and specify the location of electrons and of holes.
   (c) Suppose that $\Delta U\left(\frac{\pi}{a}, 0\right) = 2$ eV, $\Delta U\left(0, \frac{\pi}{b}\right) = 1$ eV, and that $\Delta U\left(\frac{\pi}{a}, \frac{\pi}{b}\right) = 1.5$ eV. The material is a conductor or an insulator?

(d) Consider the situation where atoms A and B are identical. Represent the new shape of the Brillouin zone.

8. Consider the transition of an electron from a fully occupied band to an empty conduction band by absorbing a photon, resulting in a hole h in the valence band. The dynamics of the valence band can be adequately described in terms of the dynamics of this pseudo-particle. Show that:

   (a) $\mathbf{k}_h = -\mathbf{k}_e$
   (b) $E_h(\mathbf{k}_h) = -E_e(\mathbf{k}_e)$
   (c) $\mathbf{v}_h(\mathbf{k}_h) = \mathbf{v}_e(\mathbf{k}_e)$
   (d) $m_h^*(\mathbf{k}_h) = -m_e^*(\mathbf{k}_e)$
   (e) under the simultaneous action of an electric field and a magnetic field, hole moves in reciprocal space according to the law

$$\hbar\left(\frac{d\mathbf{k}}{dt}\right)_h = e\left(\mathbf{E} + \mathbf{v}_h(\mathbf{k}_h) \times \mathbf{B}\right) \tag{1.84}$$

9. (a) Suppose that an electric field $\mathcal{E}$ is applied, at the time instant $t = 0$, to an electron at the bottom of a band $E(k)$ of a one-dimensional crystal. Show that the motion of the electron in real space is described by:

$$x = x_0 + \frac{E\left(k = -e\mathcal{E}t/\hbar\right)}{-e\mathcal{E}}$$

   (b) Consider a sinusoidal band as in eq. 1.83. Show that the current in real space oscillates and determine the frequency of the oscillation (these oscillations are known as *Bloch oscillations*).

10. *Kronig-Penney model*

    Consider the one-dimensional periodic potential shown in figure, where

$$U(x) = \begin{cases} 0 & , \quad 0 < x < a \\ U_0 & , \quad -b < x < 0 \end{cases}$$

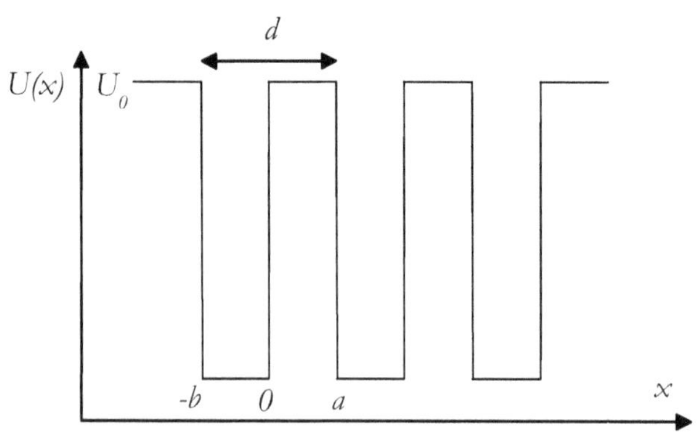

(a) Given Bloch theorem, check that the wave function obeys

$$\psi(x+d) = \exp(ik_x d)\psi(x)$$

where $d = a + b$.

(b) In the region $-b < x < a$, the wave function has the form:

$$\psi(x) = \begin{cases} Ae^{i\beta x} + Be^{-i\beta x} & , \quad -b < x < 0 \\ De^{i\alpha x} + Fe^{-i\alpha x} & , \quad 0 < x < a \end{cases}$$

where

$$\beta = \sqrt{\frac{2m_0(E-U_0)}{\hbar^2}}$$

$$\alpha = \sqrt{\frac{2m_0 E}{\hbar^2}}$$

Given the above, write the corresponding expression for the wave function in the region $a < x < a + d$.

(c) From the continuity conditions of the wave function and its derivative at $x = 0$ and $x = a$, check that the constants obey the conditions:

$$\begin{aligned} A + B &= D + F \\ \beta(A - B) &= \alpha(D - F) \\ e^{ik_x d}(Ae^{-i\beta b} + Be^{i\beta b}) &= De^{i\alpha a} + Fe^{-i\alpha a} \\ \beta e^{ik_x d}(Ae^{-i\beta b} - Be^{i\beta b}) &= \alpha(De^{i\alpha a} - Fe^{-i\alpha a}) \end{aligned}$$

(d) According to the fact that the non-trivial solution of the previous system, written in matrix form, is obtained by imposing that the determinant of the matrix of coefficients is null, check that these coefficients obey the conditions:

$$\cos(k_x d) = \begin{cases} \cos(a\alpha)\cosh(b\delta) - \frac{\alpha^2-\delta^2}{2\alpha\delta}\sin(a\alpha)\sinh(b\delta) = f(E)\,,\ 0 < E < U_0 \\ \cos(a\alpha)\cos(b\beta) - \frac{\alpha^2+\beta^2}{2\alpha\beta}\sin(a\alpha)\sin(b\beta)\,,\ E > U_0 \end{cases}$$

where

$$\delta = \sqrt{\frac{2m_0(U_0 - E)}{\hbar^2}}$$

(e) Consider the case $E < U_0$. Write a small program to plot the function $f(E)$ corresponding to the right-hand side of the previous equation as a function of E (assume reasonable values for the parameters). Given that $-1 \leq \cos(kx_d) = f(E) \leq 1$, clearly indicate the allowed energy bands on the graph.

(f) Since the equation $\cos(k_x d) = f(E)$ allows you to calculate the value of k_x for a given energy E, use a program again and plot E as a function of k_x (include at least the first three bands in the graph).

2

Semiconductor statistics

2.1 Statistics of charge carriers in solids

As we have seen, the particular properties of semiconductor materials are based on the existence of a bandgap of forbidden energies in the band structure, which separates the last occupied band (valence band) from the first unoccupied band (conduction band). Direct interband excitation of electrons from the valence band to the conduction band can occur either by optical excitation (as we shall see in the next chapter) or by thermal excitation. The latter process occurs when the thermal energy k_BT is on the order of the bandgap E_g. In the particularly important case of silicon, $E_g = 1.17$ eV and intrinsic thermal excitation occurs only at temperatures well above 1000 K. It is therefore misleading to say that semiconductors are materials of reduced bandgap. What distinguishes semiconductor materials is the ability to generate appreciable concentrations of charge carriers (electrons in the conduction band, electron holes in the valence band) at room temperature, not so much because of direct interband excitation, but essentially because of the contribution of defects and impurities present in the material. Semiconductor physics is therefore strongly intertwined with the physics of defects and impurities in semiconductors.

In order to describe the concentration n of electrons in the conduction band and the concentration p of electron holes in the valence band, relating them to the relevant physical parameters, we first write the general expressions:

$$n = \int_{E_C}^{+\infty} f_e(E)g_e(E)dE \tag{2.1}$$

$$p = \int_{-\infty}^{E_V} f_h(E)g_h(E)dE \tag{2.2}$$

where $g_e(E)dE$ and $g_h(E)dE$ are the respective densities of states; $f_e(E)$ and $f_h(E)$ are the respective probabilities of occupation. Note that, as we have seen in the previous chapter, $E_h(\mathbf{k_h}) = -E_e(\mathbf{k_e})$, where $\mathbf{k_h} = -\mathbf{k_e}$. In the above and subsequent equations, we are writing all the expressions as a function of

DOI: 10.1201/9781003587736-2

electron energies, where $E \equiv E_e$. For parabolic bands (eq. 1.25):

$$g_e(E) = \frac{\sqrt{2}}{\pi^2}\frac{m_e^{3/2}}{\hbar^3}\sqrt{E - E_C} \tag{2.3}$$

$$g_h(E) = \frac{\sqrt{2}}{\pi^2}\frac{m_h^{3/2}}{\hbar^3}\sqrt{E_V - E} \tag{2.4}$$

In eq. 2.1, $f_e(E)$ corresponds to the Fermi-Dirac function given by eq. 1.40. Furthermore, $f_h(E)$ is the probability of an electron *not occupying* the energy state E, i.e.:[1]

$$f_h(E) = 1 - f_e(E) = \frac{1}{\exp\left(\frac{\mu - E}{kT}\right) + 1} \tag{2.5}$$

If we assume the particularly important case of the non-degenerate semiconductor, where $E_C - \mu >> kT$ and $\mu - E_V >> kT$, the occupation probabilities can be written as simple Boltzmann-like functions:

$$f_e(E) \sim \exp\left(-\frac{E - \mu}{kT}\right) \tag{2.6}$$

$$f_h(E) \sim \exp\left(-\frac{\mu - E}{kT}\right) \tag{2.7}$$

The concentration of electrons in the conduction band therefore becomes

$$n \sim \int_{E_C}^{\infty} \exp\left(-\frac{E - \mu}{kT}\right) \frac{1}{2\pi^2}\left(\frac{2m_e}{\hbar}\right)^{3/2} \sqrt{E - E_C}\, dE \tag{2.8}$$

Defining $E' = E - E_C$, we may rewrite eq. 2.8 as

$$n \sim \frac{1}{2\pi^2}\left(\frac{2m_e}{\hbar}\right)^{3/2} \exp\left(\frac{\mu - E_C}{kT}\right) \int_0^{\infty} \sqrt{E'} \exp\left(-\frac{E'}{kT}\right) dE' \tag{2.9}$$

The integral in eq. 2.9 can be calculated by the substitution $y^2 = E'/kT$, resulting in $2(kT)^{3/2}\int_0^{\infty} y^2 \exp(-y^2)dy$, which can be expressed as $-2(kT)^{3/2}\frac{\partial}{\partial a}\int_0^{\infty}\exp(-ay^2)dy|_{(a=1)} = \sqrt{\pi}(kT)^{3/2}/2$. The corresponding result is:

$$n \sim 2\left(\frac{m_e kT}{2\pi\hbar^2}\right)^{3/2} \exp\left(\frac{\mu - E_C}{kT}\right) \tag{2.10}$$

Similarly, the concentration of holes in the valence band, p, for the non-degenerate case, is:

$$p \sim 2\left(\frac{m_h kT}{2\pi\hbar^2}\right)^{3/2} \exp\left(\frac{E_V - \mu}{kT}\right) \tag{2.11}$$

[1] Cf. J. S. Blakemore, *Semiconductor Statistics*, p. 81

The product np is independent of the chemical potential μ:

$$np = 4\left(\frac{kT}{2\pi\hbar^2}\right)^3 (m_e m_h)^{3/2} \exp\left(-\frac{E_g}{kT}\right) \tag{2.12}$$

where $E_g = E_C - E_V$ is the value of the range of forbidden energies between the maximum energy E_V of the valence band and the minimum energy E_C of the conduction band (bandgap).

Expression 2.12 reflects the balance between electrons in the conduction band and holes in the valence band, associated with the recombination of electrons and holes with emission of a photon γ with the bandgap energy:

$$e^- + h^+ \leftrightarrow \gamma \tag{2.13}$$

According to the statistical theory of chemical equilibrium between various chemical species, eq. 2.12 is called *law of mass action.*

2.1.1 Some limitations

Equations 2.10 and 2.11 for the concentrations of charge carriers are limited by the assumptions we have made starting from eqs. 2.1 and 2.2, in particular:

- the approximation $E_C - \mu >> kT$ and $\mu - E_V >> kT$ allowed us to consider that the occupation probability of the lowest energy electron states in the minimum of the conduction band (and of the hole states in the maximum of the valence band) is adequately described by the «tail» Boltzmann-like exponential of the respective Fermi distributions (*non-degenerate semiconductor*). If this approximation is not valid (*degenerate semiconductor*), then the full Fermi distributions have to be used in integrals 2.1 and 2.2, leading to so-called Fermi integrals without analytical solution and which have to be calculated numerically;

- we assume a parabolic band, leading to the form 1.25 for the density of states; however, it is common for bands near the minimum of the conduction band or the maximum of the valence band to have a different shape; in particular, as we saw when discussing the band structure of silicon in Chapter 1, it is common to find axially symmetrical bands with different effective masses m_l and m_t in the direction of the symmetry axis and in the perpendicular directions, respectively, where $m_e = (m_l m_t^2)^{1/3}$. In the case of valence bands, it is also common for bands with different effective masses to be degenerate: in the case where there are light holes with mass m_{lh} and heavy holes with mass m_{hh}, the effective mass m_h of the holes in the valence band becomes $m_h^{3/2} = m_{lh}^{3/2} + m_{hh}^{3/2}$.

- finally, the band structure itself varies with temperature, which is reflected in the temperature dependence of the respective parameters; the temperature dependence of bandgap is particularly important.

2.1.2 Intrinsic case

In an intrinsic semiconductor, the concentration n of electrons in the conduction band is equal to the concentration p of holes in the valence band:

$$n = p \tag{2.14}$$

The law of mass action therefore becomes:

$$np = n_i^2 = p_i^2 \tag{2.15}$$

where n_i, p_i represent the concentration of electrons in the conduction band and the concentration of holes p in the valence band in the intrinsic case, respectively, given by:

$$n_i = p_i = 2\left(\frac{kT}{2\pi\hbar^2}\right)^{3/2}(m_e m_h)^{3/4}\exp\left(-\frac{E_g}{2kT}\right) \tag{2.16}$$

Note that the temperature dependence appears in two factors of eq. 2.16 above:

- an exponential term in $\exp\left(-\frac{E_g}{2kT}\right)$, which reflects a strong dependence on temperature and a process with activation energy $E_g/2$;
- a term proportional to $T^{3/2}$, which reflects a weak dependence on temperature and is often neglected in the first approximation.

Neglecting the weak dependence on temperature, it is common to represent $\ln(n_i)$ and $ln(p_i)$ (and related quantities, such as conductivity) as a function of T^{-1} (usually as a function of $1000/T$), resulting in an approximately linear dependence. Figure 2.1 shows the concentration n_i of electrons in the conduction band of silicon, as a function of temperature, from eq. 2.16 (including the $T^{3/2}$ dependence). In the left side of figure 2.1, the n_i concentration is represented on a linear scale, and it is clear that this representation is not adequate, since it only allows to visualize the high-temperature behavior (above 800 K). In the right side of figure 2.1, a semi-logarithmic representation as a function of $1000/T$ is used, which provides a much more informative plot. This semi-logarithmic plot shows the linear behavior throughout the entire temperature range represented (from 100 K to 1000 K); it also shows that, throughout this entire temperature range, from temperatures much below room temperature to temperatures much above, the n_i concentration remains at rather low values, below 10^{20} m^{-3} (see problem 4).

Note that eq. 2.16 represents an Arrhenius-like process associated to an effective energy $E_g/2$. At first sight, this activation energy $E_g/2$ looks awkward, since the exponential increase in the concentrations n_i and p_i in eq. 2.16 is associated with the promotion of an electron from the top of the valence band to the bottom of the conduction band, therefore with an energy change E_g.

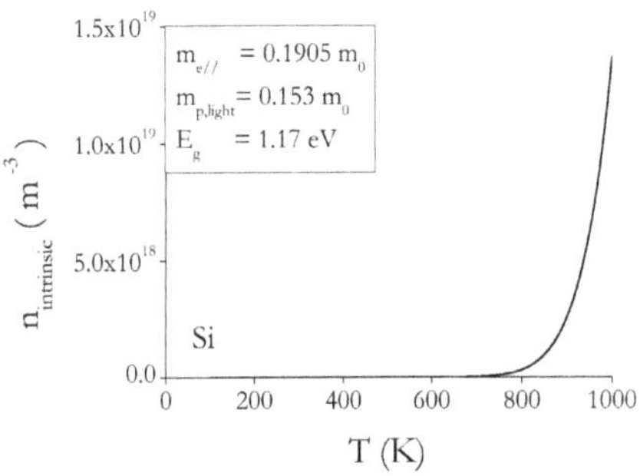

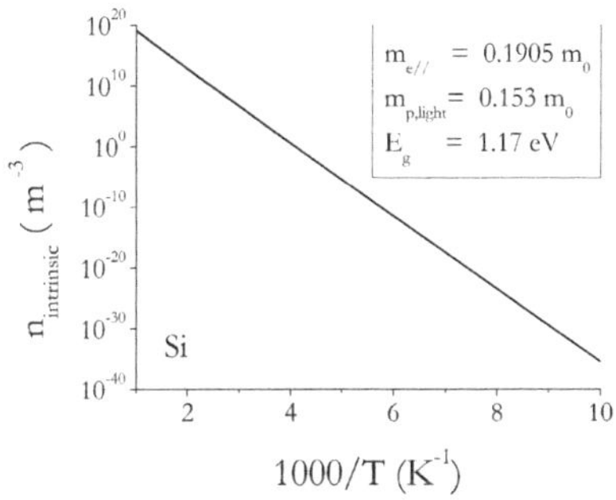

FIGURE 2.1
Concentration n_i of electrons n in the conduction band of silicon, as a function of temperature, from the eq. 2.16 (including the $T^{3/2}$ factor). **Left:** Linear scale representation as a function of temperature T; **Right:** Conventional representation on a logarithmic scale as a function of $1000/T$.

However, we must not forget that eq. 2.16 represents the statistics of electrons in a system at temperature T, where the reference for the average energy gain of the system by adding an electron is the statistical quantity *chemical potential*. The chemical potential in the case of the intrinsic semiconductor can be easily determined from the general eqs. 2.10 and 2.11, using the equality $n = p$, resulting in:

$$\mu = \frac{E_C + E_V}{2} + \frac{3}{4} kT \ln \frac{m_h}{m_e} \tag{2.17}$$

Equation 2.17 indicates that at low temperatures the chemical potential assumes the value $(E_C + E_V)/2 = E_V + E_g/2$, thus coinciding with the middle of the bandgap. Furthermore, eq. 2.17 also includes the effect of the band structure, which shows up as the slight temperature dependent term proportional to $\ln(m_h/m_e)$. When $m_h < m_e$ (as in the example of figure 2.2), the curvature of the valence band is higher than that of the conduction band $\left(d^2E/dk^2|_{BV} > d^2E/dk^2|_{BC}\right)$, so the occupation if the higher energy states of electrons in the conduction band does not progress so steeply as the occupation of the higher energy states of holes in the valence band, thus leading to a decrease in the chemical potential with temperature.

2.1.3 General aspects of the extrinsic case

The intrinsic case becomes relevant to the properties of semiconductors only at high temperatures (typically above room temperature). At room temperature and low temperatures, the concentrations of charge carriers are determined by the concentrations of impurities and defects in the material, so the equality 2.14 ($n = p$) and the corresponding eqs. 2.16 and 2.17 for n_i, p_i and the

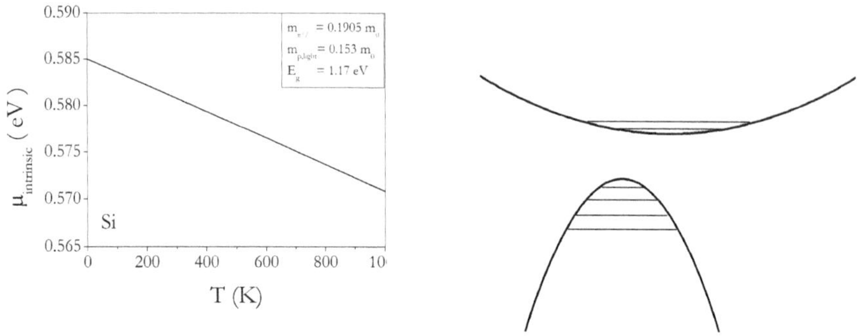

FIGURE 2.2
Left: Variation of the intrinsic chemical potential with temperature for silicon. **Right:** schematic representation of the occupation of the conduction band and the valence band, illustrating how the different curvature of the conduction band and the valence band can lead to a different occupation of the higher energy hole levels and the higher energy conduction electron levels, implying a reduction in the chemical potential with temperature as shown in the figure on the left.

chemical potential no longer apply. However, the general expressions for the concentrations of charge carriers (eqs. 2.10 and 2.11), as well as the law of mass action $np = n_i^2$ (eq. 2.12) are still valid (within the limitations mentioned before). $\Delta n = n - p$ being the difference between the concentration n of electrons in the conduction band and the concentration p of holes in the valence band, the following relations can be obtained from the law of mass action (problem 8):

$$n = \frac{\Delta n}{2} + \frac{\Delta n}{2}\sqrt{1 + \frac{4n_i^2}{\Delta n^2}} \tag{2.18}$$

$$p = -\frac{\Delta n}{2} + \frac{\Delta n}{2}\sqrt{1 + \frac{4n_i^2}{\Delta n^2}} \tag{2.19}$$

It is also possible to rewrite the general expressions 2.10 and 2.11 as a function of the intrinsic concentration n_i and p_i of charge carriers and of the corresponding chemical potential μ_i:

$$n = n_i \exp\left(\frac{\mu - \mu_i}{kT}\right) \tag{2.20}$$

$$p = p_i \exp\left(\frac{\mu_i - \mu}{kT}\right) \tag{2.21}$$

The relation between 2.20 and 2.21 thus follows:

$$\frac{\Delta n}{n_i} = 2\sinh\left(\frac{\mu - \mu_i}{kT}\right) \tag{2.22}$$

or, inverting eq. 2.22,

$$\mu = \mu_i + kT\text{arcsinh}\left(\frac{\Delta n}{2n_i}\right) \tag{2.23}$$

2.2 Defects and impurities in semiconductors

At room temperature and lower temperatures, the properties of semiconductors are dominated by the presence of impurities (atoms foreign to the crystal lattice) and defects (changes in the crystal lattice compared to the ideal lattice), so the physics of defects and impurities is intertwined with the physics of semiconductors. Understanding the physics of defects and impurities in semiconductors continues to be an extremely active research topic,[2] not only from a fundamental point of view but also from a technological point of view, since the performance of semiconductor devices is essentially conditioned by this aspect.

Defects are usually divided into *point defects* (affecting just one atom in the crystal lattice) and *extensive defects* (affecting a number of atoms in the lattice). The most important point defects are *atomic vacancies* and *antisites*. Atomic vacancies correspond to the absence of an atom from the ideal lattice site; the absence of atom A from the lattice is usually represented by V_A; antisites correspond to the case where an atom in the lattice occupies the place of another atom; the case where an atom B occupies the place of atom A is usually represented by B_A.

Extensive defects (which include phenomena such as grain boundaries, where the orientation of the lattice changes abruptly, or dislocations) are naturally much more complex and beyond the scope of this short introduction, despite their extremely important role in the properties of semiconductors.

Isolated impurities (i.e. those that do not interact with other impurities/defects) can take on two configurations in the crystal lattice: *substitutional* impurities occupy the site of an atom in the crystal lattice, replacing it; *interstitial* impurities occupy an interstitial site in the crystal lattice.

2.2.1 Hydrogenic or effective mass model

A particularly important case occurs in the case of substitutional impurities that are chemically similar to the substituted atom. The different structure of the valence orbitals of the impurity compared to the structure of the valence orbitals of the substituted atom does however have a local impact that we now present.

[2] The international conference expressly dedicated to this topic is held every two years and its 33rd edition took place in 2025.

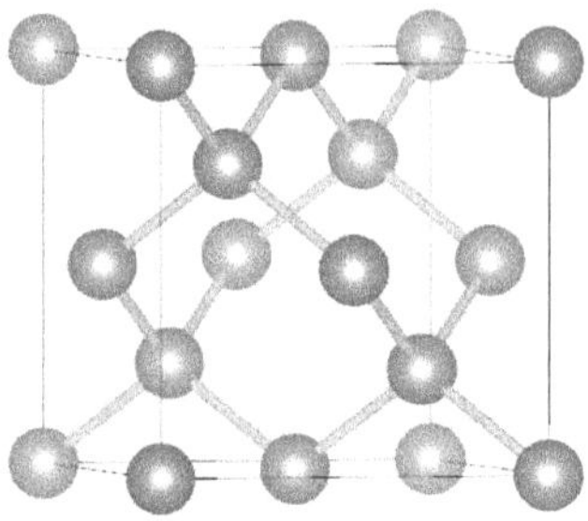

FIGURE 2.3
Crystal structure of silicon. The software VESTA was used (see text).

2.2.1.1 Classic example of a donor: phosphorus in silicon

Consider the classic example[3] of a substitutional phosphorus impurity in a silicon crystal, P_{Si}. Each atom in the silicon lattice has four valence electrons, which translates into the characteristic tetrahedral coordination of the diamond-like structure: each atom in the lattice binds to four other atoms, which form a tetrahedron around it (figure 2.3, where the VESTA software was used[4]). However, the phosphorus impurity has five valence electrons: four of the valence electrons of this impurity are used to complete the valence bonds to the surrounding silicon atoms (and therefore to complete the valence band states), but the fifth electron is in excess. Using the description of the pure Si lattice, this electron would occupy a state in the conduction band. However, at low temperatures, this fifth electron maintains its binding to the phosphorus impurity through the Coulomb[5] potential created by the remaining P^+ ion, so the corresponding energy level will be in the bandgap, placed E_b below the minimum of the conduction band (E_b being the binding energy of the fifth electron to P^+).

The binding of the fifth electron to the phosphorus impurity has two contributions: on the one hand, the potential of the crystal lattice where four valence electrons from the phosphorus atom are used to complete the valence band, forming a positively charged P^+ ion; on the other hand, the potential of the electrostatic attraction between this P^+ ion and the fifth electron. The binding energy can be easily estimated using Bohr model of a hydrogen atom, modified by the following considerations:

1. the fifth electron behaves, in practice, like a conduction electron attached to the P^+ ion;

[3] G. L. Pearson and J. Bardeen, *Electrical Properties of Pure Silicon and Silicon Alloys Containing Boron and Phosphorus*, Physical Review **75**, 865 (1949).

[4] K. Momma and F. Izumi, *VESTA 3 for three-dimensional visualization of crystal, volumetric and morphology data*, J. Appl. Crystallogr. **44**, 1272-1276 (2011).

[5] Charles-Augustin de Coulomb (1736-1806).

2. the effect of the crystal potential implies that this electron has the effective mass of the conduction electrons in the silicon lattice;
3. in addition, the attractive Coulomb potential between the P^+ ion and the electron is strongly diminished in relation to its value in vacuum due to the shielding of the charges in the material, as described by the dielectric constant of the material.

In this *modified Bohr model*, the fifth electron is subject to the following force in the presence of the P^+ ion:

$$f = \frac{1}{4\pi\epsilon}\frac{e^2}{r^2} = m^* a = m^* \frac{v^2}{r} \tag{2.24}$$

In Bohr model, angular momentum is quantized in units of $\hbar$:

$$L_n = n\hbar = m^* r v \rightarrow v = \frac{n\hbar}{m^* r} \tag{2.25}$$

Equations 2.24 and 2.25 imply that:

$$r = \frac{e^2}{4\pi\epsilon m^* v^2} = \frac{e^2 (m^*)^2 r^2}{4\pi\epsilon m^* n^2 \hbar^2} \rightarrow r_n = n^2 \frac{4\pi\epsilon\hbar^2}{e^2 m^*} \tag{2.26}$$

This equation can be rewritten as a function of the corresponding value for the hydrogen atom ($r_0 = 4\pi\epsilon_0\hbar^2/(e^2 m_0) \approx 0.53$ Å):

$$r_n = n^2 \frac{\epsilon}{\epsilon_0}\frac{m_0}{m^*} r_0 \tag{2.27}$$

In the case of silicon, $\epsilon = 11.98\,\epsilon_0$ and the effective masses of the conduction electrons, which are highly anisotropic, can reach values as low as $0.19\,m_0$: the combination of these two factors implies that the Bohr radius of the impurity can be almost two orders of magnitude larger than the value in vacuum.

The binding energy can now be easily estimated:

$$E = T + V = \frac{1}{2} m^* v^2 - \frac{1}{4\pi\epsilon}\frac{e^2}{r} = -\frac{e^4\, m^*}{2(4\pi\epsilon)^2 n^2 \hbar^2} \tag{2.28}$$

Once again, this energy can be rewritten as a function of the corresponding value for the hydrogen atom $[E_0 = -e^4\, m_0 / \left(2(4\pi\epsilon_0)^2\hbar^2\right) = -13.6 \text{ eV}]$:

$$E_n = \frac{1}{n^2}\frac{\epsilon_0^2}{\epsilon^2}\frac{m^*}{m_0} E_0 \tag{2.29}$$

In the case of silicon, the binding energy can reach values as low as c. 10 meV, which is only a fraction of the thermal energy at room temperature (c. 25 meV). This extremely low value means that the electron is very weakly bound to the ion and is easily ionized at low temperatures, moving then freely through the semiconductor like any conduction electron. The energy level of the electron bound to the impurity can therefore be represented as an isolated level about 10 meV below the minimum of the conduction band, as shown in figure 2.4 (left).

The phosphorus impurity in silicon thus provides an electron for the conduction band after ionization, which is why phosphorus is called an *impurity donor* or simply a *donor*. These weakly bound levels are usually called *shallow levels*.

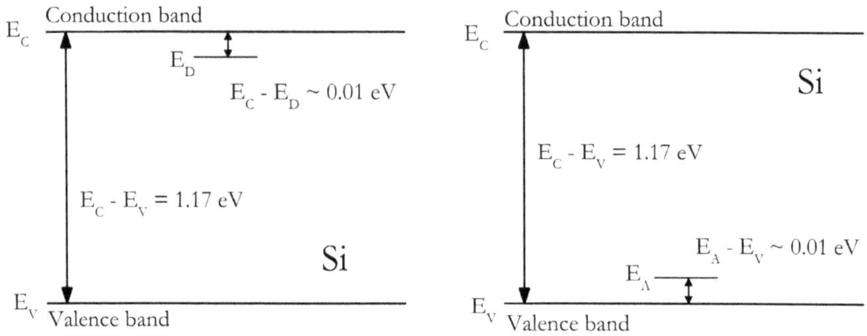

FIGURE 2.4
Left: Schematic representation of a silicon donor level. **Right:** Schematic representation of an acceptor level in silicon.

2.2.1.2 Boron in silicon: example of an acceptor

Another classic example is the case of a substitutional boron impurity in silicon. The boron atom has only 3 valence electrons, but four electrons are needed to complete the valence bonds to the surrounding silicon atoms (and therefore to complete the valence band states). Thus, a bond state is unoccupied, corresponding to an electronic hole in the valence band. This hole can be occupied by another electron in the valence band. This can be described as the movement of the hole in the valence band, leaving the boron atom negatively charged. However, the hole remains subject to the attractive potential of the B^- ion thus formed.

The modified Bohr model can also be used to estimate the binding energy of the hole to the B^- ion. In this case, we have to take into account the effective mass of the hole, which reflects the effect of the potential of the periodic lattice on the hole dynamics.

In the case of silicon, the effective mass of the holes can be as low as $0.16\,m_0$, so the binding energy can reach values as low as c. 10 meV. These extremely low values mean that the hole is weakly bound to the ion and is easily released at low temperatures, moving freely through the semiconductor. The ionization of the boron impurity thus corresponds to the capture of an electron from the valence band by the impurity or, if we prefer to phrase it in terms of holes, it corresponds to the release of a hole from the impurity into the valence band. The energy level of the hole (corresponding to an electron captured in the impurity) can therefore be represented as an isolated level about 10 meV above the maximum of the valence band, as shown in figure 2.4.

The positioning of this energy level above the maximum of the valence band correctly describes that a small energy is required to attach a valence electron to the impurity. Taking into account that the energy of the holes is symmetrical of the energy of the respective electron levels, the position of the energy level above the maximum of the valence band naturally also reflects the small energy needed to «excite the impurity hole into the valence band», in the peculiar language of holes.

In summary, after being ionized, the boron impurity in silicon captures an electron from the valence band (or «provides a hole for the valence band»), so boron is called an *acceptor impurity* or simply an *acceptor*. Again, the binding energy of the hole to the acceptor impurity can be described, in the case of substitutional impurities, by the modified Bohr model (also called hydrogenic model or effective mass model).

2.2.1.3 Beyond silicon

As we have just seen, silicon is a hallmark of semiconducting materials. Its fourfold bonding arises from the four valence electrons of the silicon atom, which granted this atom a place in the periodic table group formerly called IVA.[6] This fourfold bonding occurs in many other compound semiconductors, although with a different ionic character, depending on the group of the constituent elements. We thus have important families of semiconductors, such as the III–V family (with elements from the former IIIA and VA groups, such as GaN or GaAs) or the II–VI family (with elements from the former IIB and VIA groups, such as CdS or ZnO).

2.2.1.4 A particularly important case: atomic vacancies in partially ionic compounds

In the growth and manufacture of semiconductor materials, particularly in a laboratorial or industrial context, impurities are naturally incorporated either because they are already present in the original materials or because of the handling and manufacturing processes; the latter lead to the incorporation, sometimes in abundant concentration, of atoms associated with hydrocarbons

[6]This group nomenclature was later found to be overall ambiguous and confusing, so that it was replaced by a new nomenclature where the groups in the periodic table were simply labeled numerically from 1 to 18. The old group IVA now corresponds to group 14 in the periodic table. However, the old naming scheme is still commonly used in the context of semiconductor physics.

(H, C and O), silicates (Si), metal alloys (Fe, Cu) or simply the atmosphere (N). In addition to these particularly common impurities, the growth of compound semiconductors is also affected by inevitable small differences in the initial quantities in relation to the perfect stoichiometry, which are the cause of abundant defects. Among these, the most common defect is the absence of an atom in its crystallographic position: the *atomic vacancy.*[7] In compound semiconductors, differences in the electronegativity of the atoms lead to a different degree of ionicity of the bonds. The absence of an atom therefore impacts the electrical properties of semiconductors.

As a particularly relevant example, we consider ZnO, one of the most important compound semiconductors. ZnO is a binary semiconductor (i.e. composed of two different atoms) of the II–VI family, which crystallizes in a hexagonal structure where each atom is in tetrahedral coordination. The bonds have an ionic character, with the zinc atoms doubly positively charged and the oxygen atoms doubly negatively charged ($Zn^{2+}O^{2-}$). We start by analyzing the effect of an oxygen vacancy, V_O. The oxygen atom binds to four zinc atoms, resulting in the accumulation of two additional electrons at the position of the oxygen atom. The electrostatic potential U to which these additional electrons are subjected can be separated into two terms: the term due to the oxygen atom itself (U_O) and another term due to all the other atoms in the lattice (U_{lattice}):

$$U = U_O + U_{\text{lattice}} \tag{2.30}$$

In the absence of the oxygen atom, the potential due to the crystal lattice remains basically unchanged (at least as long as the concentration of oxygen vacancies remains low), continuing to support the location of two additional electrons on the now empty oxygen site. However, the absence of the oxygen atom potential implies that the additional electrons are now much more weakly bound to that site (the oxygen vacancy, V_O). The energy required to remove these electrons from the oxygen vacancy and promote them to the conduction band is very low and can be described by the modified Bohr model. The oxygen vacancy thus behaves like a (double) donor in ZnO. Similarly, anionic vacancies tend to be donors in the respective compounds.

An inverse effect occurs for the zinc vacancy, V_{Zn}. The zinc atom binds to four oxygen atoms, providing two electrons to form the bonds that make up the valence band. This process can be described, in the language of electronic holes, as the creation of two holes in the position of the zinc atom. These two holes remain in place because of the joint action of the lattice potential and the zinc atom potential. Again, in the absence of the zinc atom, the potential due to the crystal lattice remains basically unchanged for low concentrations of zinc atomic vacancies V_{Zn}, continuing to support the location of the two holes in that position. However, the absence of the potential due to the zinc

[7]Not to be confused with the electronic *hole*, previously discussed, corresponding to an unoccupied state in the valence band!

atom implies that these holes are now very weakly bound to the zinc vacancy V_{Zn}. The energy required to remove the electronic holes from the zinc vacancy, promoting them to the valence band, is very low, and it can be adequately described once again by the modified Bohr model. The zinc vacancy thus behaves as a (double) acceptor in ZnO and, similarly, the cationic vacancies tend to be acceptors in the respective compounds.

2.2.2 Deep impurities

As we saw in the previous section, the modified Bohr model is particularly suitable for describing the behavior of substitutional impurities or atomic vacancies. In these cases, the effect of the impurity or of the defect potential extends typically over many unit cells. However, there are many other defects or impurities that are not well described in this model. This occurs when the potential associated with these defects is quite localized, as in the case of interstitial impurities (i.e. outside a crystallographic site), or when there is a significant difference in electronegativity between the impurity and the atoms in the crystal lattice. This type of impurity or defect often has binding energies much higher than those predicted by the modified Bohr model, as well as much higher than the thermal energy $k_B T$. The corresponding energy levels are therefore further away from the ends of the bands and are called *deep levels*. As we will see in Chapter 4, these levels play an important role as centers for the recombination of conduction electrons and holes, and are also called *capture centers* or *traps*.

2.2.3 Amphoteric impurities and isoelectronic impurities

Some impurities can take on the role of either donors or acceptors, and are called *amphoteric impurities*. An example is the silicon impurity in GaAs: the silicon atom acts as a donor when it replaces a gallium atom (Si_{Ga}) and acts as an acceptor when it replaces an arsenic atom (Si_{As}). Hydrogen is also generally an amphoteric impurity.

Another relevant case is that of *isoelectronic impurities*, for example the silicon impurity in germanium (Si_{Ge}), which have no electrical activity.

2.2.4 Hydrogen: a particularly important impurity

Among the impurities present in laboratory-grown semiconductors, the atoms that make up the materials that can be found in laboratories themselves are particularly important, such as nitrogen, oxygen, hydrogen, carbon, silicon and some of the most common metals (such as iron or copper). These impurities are not all equally incorporated in the course of the growth processes. Hydrogen is a particularly ubiquitous impurity: as a constituent of hydrocarbons, it is present in the form of H^+ (protons) in combustion processes, and the small size of protons facilitates their incorporation into all materials, so hydrogen becomes an ubiquitous impurity (frequently at an appreciable concentration of up to almost 1%).

In silicon, germanium, diamond and III–V semiconductors, hydrogen tends to be an amphoteric impurity: in n-type semiconductors, it tends to occupy interstitial positions, forming deep acceptor states; in p-type semiconductors, it forms donor states (which can be deep or not) typically attached to anions in compound semiconductors or, in elementary semiconductors, even at the middle of a bond. Hydrogen, as we will see in the next section, is also an important source of compensation in most semiconductors.[8]

In some II–VI semiconductors, such as ZnO, the donor configuration (bound to an anion) is always the most stable, making hydrogen a possible source of n-type conductivity in these materials.

In addition to its behavior as an isolated impurity, hydrogen is also highly mobile inside semiconductors, rapidly forming complexes with other impurities. As we shall discuss in a moment, it therefore becomes an important passivation agent for electrically active impurities.

2.2.5 Compensation, passivation and activation

Semiconductors generally have various types of impurities and defects, with different and often opposing effects. In the simplest case where the defects and/or impurities remain isolated from each other (which is generally plausible in the low-concentration situation common in elementary semiconductors, which can be grown with a high degree of purity and crystallinity), the donor defects contribute to the total concentration of donors and the acceptor defects contribute to the total concentration of acceptors. However, the conduction electrons provided by the donors recombine with the holes provided by the acceptors, and the character of the particular sample of semiconductor material is determined by the dominant concentration: if there is an excess of donors, the semiconductor is called n-type and if there is an excess of acceptors, the semiconductor is called p-type. Note that in the simultaneous presence of isolated donors and acceptors, the concentrations of charge carriers (conduction electrons or holes) will reflect this *compensation* effect.

There are cases where defects interact and do not act isolated. The joint defect must now be taken into account, which generally occurs by adopting the perspective of the most abundant defect (typically introduced deliberately, as will be discussed below). We therefore speak of the *passivation* of a donor or acceptor defect in the case where the joint defect does not have electrical activity (i.e. it does not introduce a level of its own into the bandgap); it is also common to say that the less abundant defect «removed» the level of the more abundant defect (donor or acceptor) from the bandgap. A classic example

[8] Cf. R. C. Vilão, in *Muon Spectroscopy: an introduction*, pp. 157-169.

is the removal of the acceptor levels of magnesium impurities deliberately introduced into GaN, due to the formation of an Mg-H complex.[9]

The opposite case, called *activation*, also occurs, although it is less important: it can happen that an impurity or defect does not show electrical activity *per se*, but only when it establishes a joint defect with another defect/impurity. This is the case, for example, with the isoelectronic impurities of silicon in germanium, which show acceptor behavior after forming a Si-H complex.

2.2.6 Statistics of impurity levels

The presence of impurity levels in the bandgap has an extraordinarily relevant impact on the behavior of electrons in the material. To quantify the impact of these levels, it is important to determine their probability of occupation. To do this, it should be remembered that the occupation of the level implies the exchange of an electron from the impurity with the solid, carried out at the average energy μ (chemical potential of the system). The average occupation $< n >$ of the level at a given temperature T can thus be expressed as follows:[10]

$$< n >= \frac{\Sigma_j N_j \exp\left[-\frac{E_j-\mu N_j}{k_B T}\right]}{\Sigma_j \exp\left[-\frac{E_j-\mu N_j}{k_B T}\right]} \tag{2.31}$$

where N_j is the number of electrons that occupy the energy level E_j at temperature T.

In the case of a non-degenerate donor level, we have the following possibilities:

1. the impurity level may be unoccupied and the total energy is $E_1 = 0$;
2. the impurity level may be occupied by an electron with spin *up*, and the total energy is $E_1 = E_D$;
3. the impurity level can be occupied by an electron with spin *down*, and the total energy is $E_1 = E_D$;

Note that in the case where the level is doubly occupied (with an electron with spin *up* and another with spin *down*), the repulsion energy between the two electrons would have to be added; in practice, this would push the level to values much above the minimum of the conduction band, leading to immediate ionization. In practice, double occupancy can be ignored.

[9] Understanding this effect was essential for the possibility of using GaN in the manufacture of blue LEDs, the basis of the «revolution» that LEDs introduced into lighting. These investigations were awarded the Nobel Prize for Physics in 2014. Cf. Shuji Nakamura, *Nobel Lecture: Background story of the invention of efficient blue InGaN light emitting diodes*, Reviews of Modern Physics **87**, 1139 (2015).

[10] Neil D. Ashcroft and D. David Mermin, *Solid State Physics*, p. 581.

From eq. 2.31, we can now determine the average occupancy of the level:

$$< n >= \frac{2\exp\left[-\frac{E_D-\mu}{k_BT}\right]}{1+2\exp\left[-\frac{E_D-\mu}{k_BT}\right]} = \frac{1}{1+\frac{1}{2}\exp\left[+\frac{E_D-\mu}{k_BT}\right]} \tag{2.32}$$

The factor $1/2 = 1/g$ in eq. 2.32 reflects the fact that there are two possibilities for occupation of the donor level. g is thus called the *degeneracy factor* of the level. The case we have just discussed corresponds to the situation where the donor level has a s character; in cases where the donor level is formed mainly by p orbitals, the degeneracy factor will be correspondingly affected by the triple degeneracy of these orbitals, which can lead to $g = 6$ (3 p orbitals, each of which can be occupied by an electron with spin *up* and another with spin *down*). So, in general, for a given degeneracy factor g:

$$< n >= \frac{N_D^0}{N_D} = \frac{1}{1+\frac{1}{g}\exp\left[+\frac{E_D-\mu}{k_BT}\right]} \tag{2.33}$$

Equation 2.33 represents, as we have seen, the average occupancy of the donor level, or the fraction of non-ionized donors (ratio between the concentration of non-ionized donors N_D^0 and the total concentration of donors N_D). The corresponding fraction of ionized donors immediately results as:

$$\frac{N_D^+}{N_D} = 1 - \frac{N_D^0}{N_D} = \frac{1}{1+g\exp\left[-\frac{E_D-\mu}{k_BT}\right]} \tag{2.34}$$

Similarly, it is possible to obtain the average occupancy of the acceptor level, or the fraction of non-ionized acceptors (ratio between the concentration of non-ionized acceptors N_A^0 and the total concentration of donors N_A):

$$\frac{N_A^0}{N_A} = \frac{1}{1+\frac{1}{g_A}\exp\left[+\frac{\mu-E_A}{k_BT}\right]} \tag{2.35}$$

and the corresponding fraction of ionized acceptors:

$$\frac{N_A^-}{N_A} = \frac{1}{1+g_A\exp\left[-\frac{\mu-E_A}{k_BT}\right]} \tag{2.36}$$

2.2.6.1 Donor ionization

We will now consider the contribution of donor ionization to the concentration of electrons in the conduction band. We start with the simplest situation where a semiconductor has a concentration of N_D simple donors, all with the same ionization energy E_D. We assume that the concentration of donors is small enough for the donors to remain isolated. The donors are presumably (but not necessarily) described by the effective mass model, with a reduced ionization energy («*shallow*» donors). Since $E_D << E_g$, the donor ionization process for the conduction band dominates at low temperatures, long before the intrinsic interband excitation process takes place. Under these conditions, the electrons in the conduction band come essentially from ionized donors and the concentration n of conduction electrons corresponds to the concentration N_D^+ of ionized donors. The electrical neutrality condition therefore becomes:

$$n = N_D^+ \tag{2.37}$$

Given the general expression 2.10 for n and the eq. 2.34 for the fraction of ionized donors, the electrical neutrality condition can be rewritten as:

$$N_C \exp\left(\frac{\mu - E_C}{kT}\right) = \frac{N_D}{1 + g\exp\left[-\frac{E_D-\mu}{k_B T}\right]} \tag{2.38}$$

where $N_C = 2(m_e kT/2\pi\hbar^2)^{3/2} >> N_D$. This equation determines the position of the chemical potential μ as a function of temperature, which can be easily found to be (problem 11):

$$\mu = E_D + kT\ln\left[\frac{-1 + \left[1 + 4g\frac{N_D}{N_C}\exp\left(\frac{E_C-E_D}{kT}\right)\right]^{\frac{1}{2}}}{2g}\right] \tag{2.39}$$

The corresponding concentration of electrons in the conduction band is then given by:

$$n = \frac{2N_D}{1 + \left[1 + 4g\frac{N_D}{N_C}\exp\left(\frac{E_C-E_D}{kT}\right)\right]^{\frac{1}{2}}} \tag{2.40}$$

Low temperature limit

At the low temperature limit, eq. 2.39 returns, as expected, a chemical potential located halfway between the donor level position and the minimum of the conduction band:

$$\mu \sim \frac{E_C + E_D}{2} + \frac{1}{2}kT\ln\frac{N_D}{gN_C} \tag{2.41}$$

which corresponds to a concentration of electrons in the conduction band that grows exponentially with an activation energy of $(E_C - E_D)/2$:

$$n \sim \sqrt{\frac{N_C N_D}{g}}\exp\left(-\frac{E_C - E_D}{2kT}\right) \tag{2.42}$$

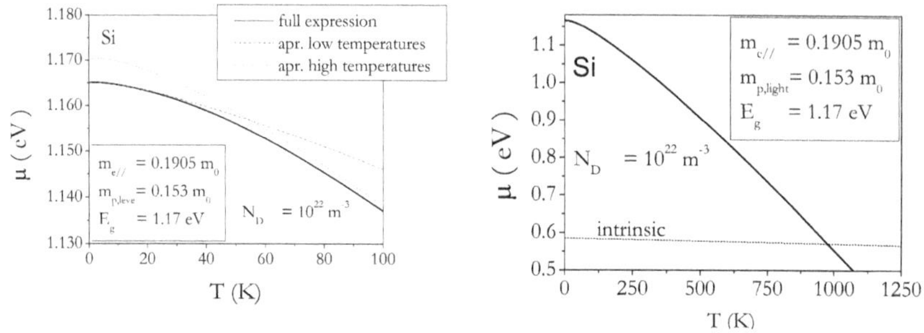

FIGURE 2.5
Variation with temperature of the chemical potential of silicon doped with hydrogenic impurities (with ionization energy of 10 meV and concentration 10^{22} m^{-3}). **Left:** Behaviour across the transition from donor freezing to donor ionization (up to T = 100 K): the solid black line represents eq. 2.39; the dashed line represents the low temperature approximation (eq. 2.41) and the dotted line represents the high temperature approximation (eq. 2.43). **Right:** extension to high temperatures up to the intrinsic regime: the solid line represents eq. 2.39 and the dashed line represents the intrinsic chemical potential given by eq. 2.17.

High temperature limit

In the high temperature limit (which corresponds to the saturation regime where all the donors are ionized), the chemical potential is approximately described by the expression:

$$\mu \sim E_C + kT \ln \frac{N_D}{N_C} \tag{2.43}$$

which corresponds to a concentration of electrons in the conduction band identical to the concentration of donors ($n \sim N_D$).

$$n \sim N_D \tag{2.44}$$

Figures 2.5 and 2.6 illustrate the temperature dependence of the chemical potential and of the electron concentration n in the conduction band, given by the model of eqs. 2.39 and 2.40.

2.2.6.2 Including the intrinsic regime

In the previous section, we have described the particularly important case of donor ionization. For shallow impurities, this occurs at temperatures typically below room temperature and the concentration of conduction electrons quickly saturates at the donor concentration N_D. In general, it is only at much higher temperatures (typically above room temperature) that the direct process of interband excitation becomes apparent. In this case, the

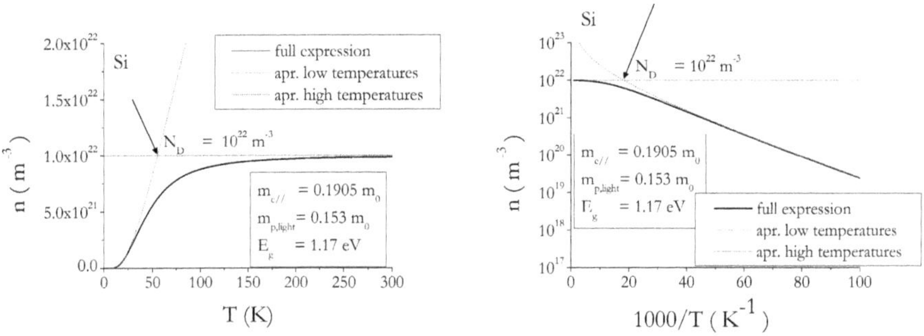

FIGURE 2.6
Concentration of electrons n in the conduction band of silicon doped with hydrogenic impurities (with ionization energy of 10 meV and concentration 10^{22} m^{-3}). On the left is the representation on a linear scale as a function of temperature T; on the right is the representation on a logarithmic scale as a function of $1000/T$. The solid black line represents eq. 2.40. The dashed line represents the low temperature approximation $n \sim N_D$ and the dotted line represents the low temperature approximation (eq. 2.42). The arrows indicate the intersection temperature of the two approximate curves, which provides an estimate of the ionization temperature of the donors.

intrinsic process also contributes significantly to the concentration of conduction electrons. The condition of electrical neutrality then becomes:

$$n = p + N_D^+ \tag{2.45}$$

Given the general expression 2.10 for n, eq. 2.34 for the fraction of ionized donors and the law of mass action $p = n_i^2/n$, the electrical neutrality condition can be rewritten as:

$$N_C \exp\left(\frac{\mu - E_C}{k_B T}\right) = \frac{n_i^2}{N_C \exp\left(\frac{\mu - E_C}{k_B T}\right)} + \frac{N_D}{1 + g \exp\left[-\frac{E_D - \mu}{k_B T}\right]} \tag{2.46}$$

This equation can be solved in order to provide the chemical potential:[11]

$$\mu = E_D + k_B T \ln\left[\frac{-1 + \frac{\beta}{\gamma} + \frac{\gamma}{N_C^2}}{3g}\right] \tag{2.47}$$

where

[11]Cf. M. Grundmann, *The Physics of Semiconductors*, p. 200.

$$\gamma = \left[\frac{-N_C^4\alpha + \sqrt{(N_C^4\alpha)^2 - 4(N_C^2\beta)^3}}{2}\right]^{\frac{1}{3}} \quad (2.48)$$

$$\beta = N_C^2 + 3gN_CN_D\exp\left(\frac{E_D}{k_BT}\right) + 3g^2n_i^2\exp\left(\frac{2E_D}{k_BT}\right) \quad (2.49)$$

$$\alpha = 2N_C^2 + 9gN_CN_D\exp\left(\frac{E_D}{kT}\right) - 18g^2n_i^2\exp\left(\frac{2E_D}{k_BT}\right) \quad (2.50)$$

The corresponding concentration of electrons in the conduction band is:

$$n = N_C\exp\left(-\frac{E_C - E_D}{k_BT}\right)\frac{-1 + \frac{\beta}{\gamma} + \frac{\gamma}{N_C^2}}{3g} \quad (2.51)$$

Equation 2.51 presents three important regimes:

1. intrinsic regime ($n \approx n_i >> N_D$): at high temperatures, the intrinsic regime is dominant, and the growth of n is basically governed by the factor $\exp(E_g/2kT)$, according to eq. 2.16;
2. exhaust or saturation regime ($n \approx N_D >> n_i$ and $kT > E_D$): at intermediate temperatures, the donors are completely ionized and the intrinsic regime is negligible, with the concentration of electrons in the conduction band basically equal to the concentration of donors ($n \sim N_D$) (eq. 2.44); in silicon and in most semiconductors of technological interest, this regime dominates at room temperature;
3. ionization or freeze-out regime ($kT << E_D$): at sufficiently low temperatures, the donors are partially ionized, and the concentration of electrons in the conduction band is basically governed by $\exp((E_C - E_D)/2kT)$ according to eq. 2.42;

The depletion or saturation regime, illustrated in figure 2.7, thus corresponds to a temperature range, typically around room temperature, where the electrical properties of the semiconductor are not only relatively stable, but are, above all, controllable through doping, through impurities and/or defects deliberately introduced in the growth process. This regime is therefore the regime of choice for most technological applications.

In the case of shallow impurities in silicon, the ionization or freezing regime is only relevant at temperatures much lower than room temperature, and must be taken into account in applications intended for use at low temperatures (such as space applications). In some semiconductors (cf. problem 19), this regime can be relevant at room temperature. In silicon, as figure 2.7 clearly illustrates, the intrinsic regime only becomes relevant at temperatures much higher than room temperature. In other semiconductors with a narrower bandgap (particularly Ge, which can be produced with exceptionally high purity), the intrinsic regime can become relevant at room temperature or just above.

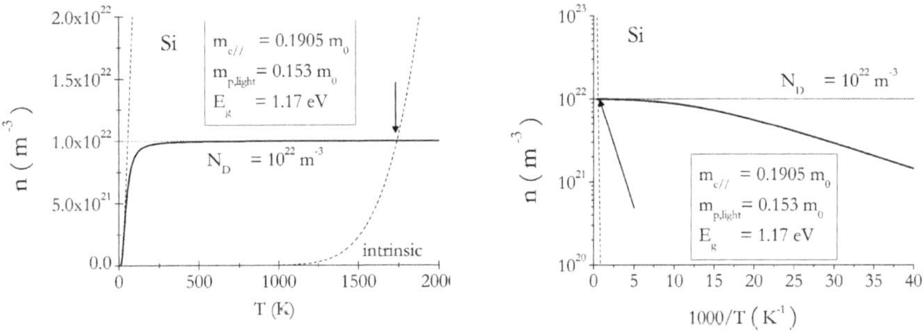

FIGURE 2.7
Concentration of electrons n in the conduction band of silicon doped with hydrogenic impurities (*shallow*, with ionization energy of 10 meV and concentration 10^{22} m^{-3}). **Left:** the linear scale representation as a function of temperature T. **Right:** conventional representation on a logarithmic scale as a function of $1000/T$. The solid black line represents eq. 2.40. The dashed line represents the low temperature approximation (eq. 2.42: ionization or *freeze-out* regime) and the dotted line represents the saturation regime (eq. 2.44). The dashed line at high temperatures represents the intrinsic regime given by eq. 2.16. The arrows indicate the intersection temperature of the saturation regime with the intrinsic regime, which provides an estimate of the temperature around which the intrinsic regime becomes dominant.

2.2.6.3 Compensation

In a realistic case, a given semiconductor has multiple electrically active defects/impurities. The situation of multiple donors or multiple acceptors can be dealt with easily. However, the usual situation where there is compensation of donors and acceptors requires special attention (problem 14). In the case of an n-type semiconductor dominated by donors of ionization energy $E_D^i = E_C - E_D$ and concentration N_D, compensated by acceptors with concentration N_A, the following expression is obtained for the concentration of electrons in the conduction band:

$$\frac{n(N_A + n)}{N_D - N_A - n} = \frac{N_C}{g} \exp\left(-\frac{E_D^i}{kT}\right) \tag{2.52}$$

2.2.6.4 High doping

Up to now, we have assumed that impurities do not significantly affect the band structure of the semiconductor. In this situation, the impurities are far enough apart that there is no appreciable overlap of their wave functions, and their energy levels remain localized. However, for high concentrations, the wave functions of the impurities can overlap significantly, particularly in the case of impurities described by the effective mass model, which correspond to high values of the Bohr radius. In this case, the impurity wave functions give rise to an *impurity band*. Consider the case of an impurity band generated by a single type of donor. At low temperatures, the impurity band is completely occupied and the semiconductor behaves like an insulator. However, if there are also acceptors in the semiconductor, compensation will occur and some of the acceptor levels will be occupied by electrons from the impurity band. At low temperatures, in a compensated semiconductor, the impurity band will be partially empty and electrical conduction can occur at these temperatures, with the semiconductor behaving like a metal. The transition from insulating behavior to metallic behavior at low temperatures occurs abruptly when the impurity concentration exceeds a threshold that allows the impurity band to form (*Mott's criterion*[12]). If there are N donors in the V volume of the sample, the impurity concentration is $N_D = N/V$ and the sample volume that contains, on average, one donor is $V_1 = V/N = 1/N_D$. For low donor concentrations, V_1 is much larger than the volume occupied by the donor wave function $V_{Bohr} = 4\pi r_B^3/3$, where r_B is the Bohr radius of the impurity. As the donor concentration increases, V_1 decreases up to a value where:

$$V_1 = \frac{1}{N_D} = \frac{4\pi r_B^3}{3} \rightarrow N_D = \frac{3}{4\pi r_B^3} \tag{2.53}$$

For values of N_D larger than those expressed in eq. 2.53, the average sample volume containing a donor is less than the volume occupied by the donor's wave function and the overlap of the wave functions becomes relevant.

Problems

1. Obtain eq. 2.11 for the concentration p of holes in the valence band.
2. Obtain eq. 2.17 for the chemical potential in the intrinsic case.
3. The pre-exponential factors (or effective densities of states) for the intrinsic electron and hole concentrations for Ge are, respectively, 1.0×10^{19} cm^{-3} and 6.0×10^{18} cm^{-3}, at room temperature (300 K). Calculate:
   (a) the effective masses of electrons and holes;
   (b) the concentration of charge carriers at 300 K and at the boiling temperature of liquid nitrogen (77 K).
4. The pre-exponential factors (or effective densities of states) for the intrinsic electron and hole concentrations for Si are, respectively,

[12] Neville Mott (1905-1996).

2.8×10^{19} cm^{-3} and 9.8×10^{18} cm^{-3}, at room temperature (300 K). The bandgap of silicon at room temperature is 1.12 eV. The density of silicon is 2330 kg cm^{-3}. Consider a sample of pure Si.

(a) Calculate the concentration of silicon atoms.

(b) Calculate the intrinsic concentration of electrons and holes at room temperature, at $T = 100$ K and at $T = 1000$ K; how many electrons/holes per atom are expected at each temperature?

(c) Determine the value of the chemical potential at the temperatures above.

5. The figure shows the resistance (on a logarithmic scale) of a germanium sample, as a function of the inverse of the temperature (1000/T). Estimate, from the graph, the value of the energy bandgap.

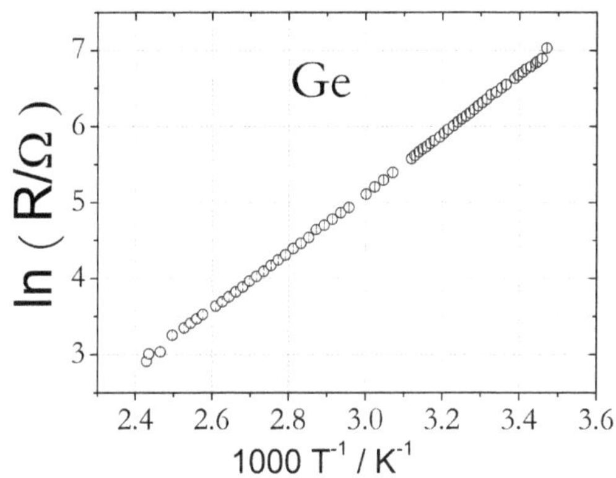

6. A sample of Si has been purified until it contains only 10^{12} donors/cm^3, of ionization energy $E_D = 45$ meV. Under these circumstances, estimate:

(a) the transition temperature to intrinsic behavior;

(b) the freeze-out temperature of donors.

7. Three samples of Ge doped with As ($E_D = 13$ meV) have, respectively, the following donor concentrations: 5.5×10^{16} cm^{-3}, 1.7×10^{15} cm^{-3} and 1.4×10^{14} cm^{-3}. In all of them, the concentration of acceptors is $N_A << 10^{13}$ cm^{-3}. Plot $n(T)$ for the three samples on a semi-logarithmic scale and comment on the types of variation observed on the graph. Classify the behavior of each of the samples at room temperature as «intrinsic» or «extrinsic».

8. Consider a doped semiconductor. Obtain eqs. 2.18, 2.19 and 2.22.

9. InSb has a band structure with $E_g = 0.236$ eV and $m_e^* = 0.015\, m_0$. The relative dielectric constant of this semiconductor is equal to

17. Assuming that a sample of InSb is doped with donor impurities adequately described by the effective-mass model, calculate:

(a) the binding energy of the donors;

(b) the radius of the electron orbit provided by the donor;

(c) the minimum concentration of donors so that there is appreciable overlap between the orbits of the electrons in adjacent impurity atoms.

10. A sample of Ge is doped with only one type of impurity. If the concentration of donors is 10^{14} cm^{-3}, what are the experimental conditions necessary to observe cyclotron resonance? ($m_e^* = m_0$; the orbit diameter of the donor electron is approximately 30 Å).

11. Consider a semiconductor doped with a concentration N_D of ionization energy donors E_D.

(a) Write the electrical neutrality condition, neglecting the contribution of intrinsic charge carriers.

(b) Determine the chemical potential μ as a function of temperature.

(c) Determine the concentration n of electrons in the conduction band as a function of temperature.

(d) Determine approximate expressions for μ and for n in the limit $T \to 0$ and in the limit of high temperatures. Sketch a graph of the temperature dependence of μ and n.

(e) How do the graphs in the previous paragraph change when the contribution of intrinsic behavior is taken into account?

(f) Write the electrical neutrality condition including the contribution of intrinsic charge carriers.

12. Repeat the previous problem for the case of a semiconductor doped with a concentration N_A of acceptors *shallow* of ionization energy E_A.

13. CdTe is a high bandgap semiconductor of the II−VI family. The bonds in CdTe are ionic in nature (Cd^{2+}−Te^{2-}).

($m_e^* = 0.09\, m_0$, $m_h^* = 0.8\, m_0$, $\epsilon_{stat} = 10.2$)

(a) Indicate impurities with possible (simple) donor behavior and with possible (simple) acceptor behavior, described by the effective mass theory; calculate the respective Bohr radii and binding energies.

(b) Indicate impurities with possible amphoteric behavior.

(c) Indicate the electrical behavior expected for the structural vacancies of Cd, V_{Cd}, and Te, V_{Te}.

14. (a) Determine the concentration of acceptor and donor impurities in a semiconductor that are ionized at temperature T. Using the electrical neutrality condition

$$p + N_D - n_D = n + N_A - p_A$$

show that

$$N_V \exp\left(-\frac{\mu}{k_B T}\right) + \frac{N_D}{1 + \frac{1}{\beta}\exp\left(-\frac{\epsilon_D - \mu}{k_B T}\right)} =$$

$$N_C \exp\left(-\frac{E_g - \mu}{k_B T}\right) + \frac{N_A}{1 + \frac{1}{\beta}\exp\left(-\frac{\mu - \epsilon_A}{k_B T}\right)}$$

where the reference level for the energies is the bottom of the valence band, $\beta = 1/2$ is the degeneracy factor of the donor and acceptor levels, N_D is the concentration of donors and N_A is the concentration of acceptors; n_D and p_A are the concentrations of non-ionized donors and acceptors, respectively; $N_C = 2\left(\frac{m_e^* k_B T}{2\pi\hbar^2}\right)^{\frac{3}{2}}$ and $N_V = 2\left(\frac{m_h^* k_B T}{2\pi\hbar^2}\right)^{\frac{3}{2}}$ are the effective densities of states in the conduction and valence bands, respectively.

(b) In order to obtain the temperature dependence of n in a compensated semiconductor of type n ($N_D > N_A$), consider p and p_A to be negligible. Show that:

$$\frac{N_D}{1 + \beta \exp\frac{E_D - \mu}{kT}} = N_D - N_A - n$$

(c) From the expression above, show that

$$\exp\frac{\mu}{kT} = \beta\frac{N_D - N_A - n}{N_A + n}\exp\frac{E_D}{kT}$$

(d) Check that

$$\frac{n(N_A + n)}{N_D - N_A - n} = \beta N_C \exp\left(-\frac{E_D^i}{kT}\right),$$

where $E_D^i = E_g - E_D$ is the ionization energy of the donor.

(e) Consider the limiting case $n << N_D, N_A$ and obtain an expression for the chemical potential, $\mu(T)$, and for the concentration $n(T)$, showing that at very low temperatures the chemical potential tends to $\mu = E_D$; repeat for the limiting case $N_A << n < N_D$, where the chemical potential tends to $\mu = \frac{E_g + E_D}{2}$.

(f) Confirm the previous results numerically by performing a simulation for a Si wafer doped with a donor concentration $N_D = 1 \times 10^{18}$ cm^{-3} and the following values for the acceptor concentration: $N_A = 0$, $N_A = 10^{13}$ cm^{-3}, $N_A = 10^{14}$ cm^{-3}, $N_A = 10^{15}$ cm^{-3}, $N_A = 10^{16}$ cm^{-3} and $N_A = 10^{17}$ cm^{-3}.

Plot the chemical potential and the concentration of charge carriers in the temperature range between 5 and 800 K. Compare your results with data in the literature.

15. (a) Using a programming language of your choice, write a program that, starting from the electrical neutrality condition (eqs. 2.45 and 2.46), numerically determines the temperature dependence of the chemical potential and the electron concentration in the conduction band (eqs. 2.47 and 2.51).

(b) Generalize the program to include compensation via an acceptor.

(c) Generalize the program to include multiple acceptors and donors.

(d) Generalize the program for the case of a degenerate semiconductor (cf. eqs. 1.40, 2.1, 2.2 and 2.5).

16. InAs is a III–V family semiconductor, with partially ionic bonds (In^{+}–As^{-}). The table shows the results, in ppm (weight), of mass spectrometry measurements on InAs, in a sample of the type n.[13] The mobility measured by Hall effect was $\mu_e = 2.3 \times 10^4$ cm^2 V^{-1} s^{-1} and the electron concentration in the conduction band was $n = 2 \times 10^{16}$ cm^{-3}.

(a) Indicate the expected electrical behavior for each of the impurities indicated in the table (except Fe, Mn and Cu), assuming that they are adequately described by the effective mass model.

(b) Calculate the respective Bohr radius and binding energy of the shallow donors and acceptors in InAs.

(c) The possibility of carbon atoms occupying substitutional positions in As seems unlikely. Assuming that only C and O are electrically active impurities, estimate the expected electron concentration in the saturation regime.

(d) One possibility to justify the difference between the result in the previous paragraph and the experimental result is compensation by structural defects. In this hypothesis, which defect should be the most abundant? Please also indicate another hypothesis to justify the experimental result.

---

[13] Hu Weijie *et al.*, *Residual impurities and electrical properties of undoped LEC InAs single crystals*, Journal of Semiconductors **31**, 042001 (2010).

Element	Content in ppm (weight)	Element	Content in ppm (weight)	Element	Content in ppm (weight)
B	0.03	K	0.02	Ca	0.01
C	23	N	0.64	Mn	0.001
O	10	Fe	0.02	Cu	0.005
Mg	0.004	Al	0.002	Zn	0.01
Si	0.01	Cd	0.02	Sn	0.02
Te	0.006	Cl	0.002	P	1.6
Sb	0.005	S	0.24		

17. Estimate the nature (donor, acceptor) of the electrical behavior of the following impurities/defects. Indicate whether you expect the wave function to be adequately described by the effective mass model (impurities *shallow*/*deep*). Also indicate the cases of double donors/acceptors, as well as the cases of amphoteric impurities.

(a) As_{Si} in Si;
(b) B_{Si} in Si;
(c) Ge_{Si} in Si;
(d) Se_{Si} in Si;
(e) Zn_{Si} in Si;
(f) Au_{int} in Si;
(g) V_{Ga} in GaAs;
(h) V_{As} in GaAs;
(i) Ge_{As} in GaAs;
(j) Ge_{Ga} in GaAs;
(k) As_{Ga} in GaAs;
(l) V_O in ZnO;
(m) V_{Zn} in ZnO;
(n) $H - O$ in ZnO;
(o) H_{int} in ZnO;

Notes on the notation:

- X_Z indicates that impurity X is in a substitutional position, replacing a Z atom in the crystal lattice;
- X_{int} indicates that impurity X is in an interstitial position, which does not correspond to any atomic position in the crystal lattice;
- V_Z indicates the absence (*vacancy*) of the Z atom;
- $H - O$ indicates the respective covalent bond.

18. Cu_2ZnSnS_4 (kesterite) is a quaternary semiconductor derived from the II–VI family, with partially ionic bonds where the metallic atoms Cu, Zn and Sn transfer electrons to the non-metallic atom S. This

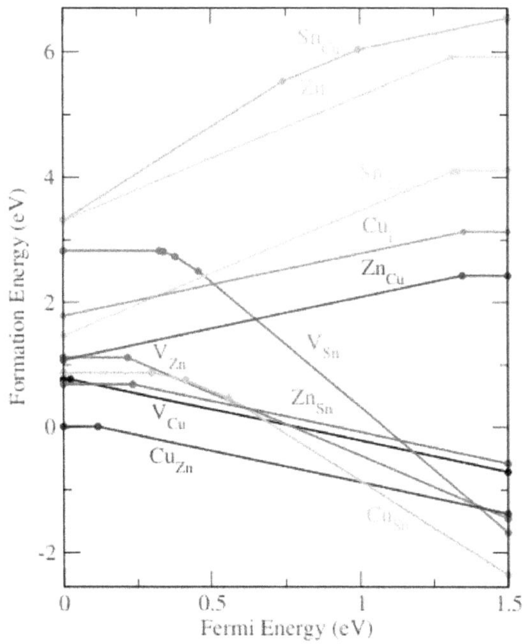

FIGURE 2.8
Formation energies of intrinsic defects in kesterite, as a function of Fermi energy (figure reproduced with permission from S. Chen, J.-H- Yang, X. G. Gong, A. Walsh, and S.-H. Wei, *Intrinsic point defects and complexes in the quaternary kesterite semiconductor Cu_2ZnSnS_4*, Phys. Rev. B **81**, 245204 (2010). *Copyright* (2010) of the American Physical Society.)

material is very difficult to grow and structural defects play a decisive role. The calculated band structure of kesterite reveals a direct bandgap ($m_e^* = 0.18\, m_0$, $m_h^* = 0.43\, m_0$, $\epsilon_r = 6.8$). The bandgap of Cu_2ZnSnS_4 is considered close to ideal for application in solar cells, not only because of its direct nature, but also because of its value $E_g = 1.5$ eV. Fig. 2.8 shows the formation energies of various structural defects as a function of the Fermi energy (chemical potential at $T = 0$ K).

(a) Indicate the electrical character (donor or acceptor) of the structural defects shown in the figure.
(b) Calculate the Bohr radii and binding energies of the defects for which you expect the effective-mass model to apply.
(c) According to the fact that the structural defects with the lowest formation energy are the most abundant, what type of conductivity (n or p) do you expect for «intrinsic» Cu_2ZnSnS_4?

(d) Estimate the temperature at which «freezing» of the impurity levels occurs, as well as the temperature at which it is no longer possible to disregard the intrinsic process.

19. GaN is a semiconductor from the III–V family, with $E_g = 3.4$ eV. p-type GaN is produced by doping with Mg. However, the Mg acceptors are passivated with hydrogen, which implies an additional step in the processing of the material in order to activate the acceptors by dissociating the Mg–H complex. This typically involves annealing at high temperatures. Figure 2.9 contains results of the concentration of electronic holes in GaN doped with Mg, annealed at various temperatures. The lines in figure 2.9 represent fits to a condition of electrical neutrality assuming a single acceptor with an ionization energy of around 180 meV, as well as partial compensation by an unidentified donor with a concentration of 3×10^{18} cm^{-3}. ($m_e = 0.2\, m_0$, $m_h = 1.25\, m_0$, $\epsilon_r = 9.7$)

    (a) Can the increase in the concentration of electronic holes with temperature be explained by the intrinsic process? Justify by giving an estimate of the concentration of holes at 300 K due only to this process, as well as an estimate of the temperature at which it becomes dominant.

    (b) Is the experimentally measured ionization energy consistent with acceptors described by the effective mass model?

    (c) Estimate the concentration of acceptors in each of the samples, assuming that the fitting lines reach the saturation value at the highest temperature represented.

    (d) Is the dissociation of the Mg–H complexes consistent with the increase in acceptor concentration with annealing temperature?

    (e) The compensating donors can be native atomic vacancies or Si atoms occupying substitutional sites. Indicate the candidate atomic vacancy as well as the candidate Si substitutional site.

    (f) The resistivity at room temperature of the sample annealed at 775°C is $2\,\Omega$cm. Estimate the mobility and diffusion constant of the electronic holes in this sample, as well as the average time between collisions for the holes.

20. Ga_2O_3 is a very promising material for applications in high-power electronics. Figure 2.10 shows measurements of the concentration of charge carriers in the conduction band, obtained through Hall effect measurements, for various Ga_2O_3 samples doped with Si or Ge donors. The lines represent fits to a condition of electrical neutrality assuming a single donor species with partial compensation with an acceptor. ($m_e = 0.3\, m_0$, $\epsilon_r = 10$)

    (a) Do you expect the Si and Ge donors to occupy Ga or O positions? Justify.

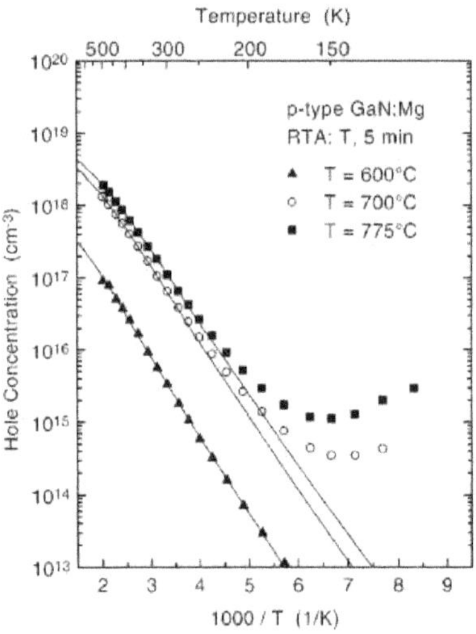

FIGURE 2.9
Experimental results of the measurement of the concentration of electronic holes in samples of GaN doped with Mg, annealed at various temperatures, obtained by W Götz, N. M. Johnson, J. Walker, D. P. Bour, and R. A. Street, *Activation of acceptors in Mg-doped GaN grown by metallorganic chemical vapor deposition*, App. Phys. Lett. **68**, 667 (1996). Reproduced with permission of the publisher AIP Publishing LLC.

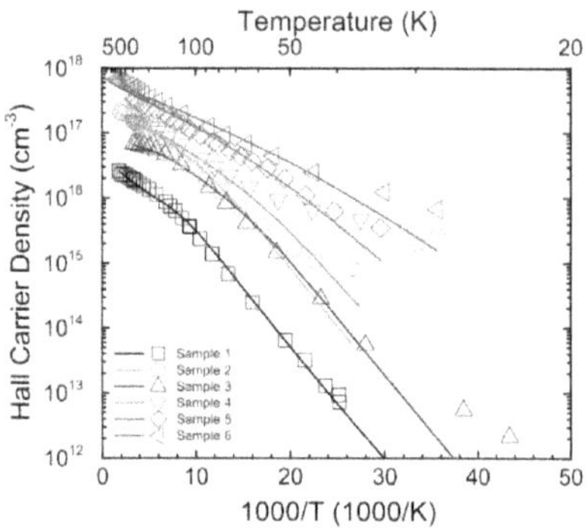

FIGURE 2.10
Measurements of the concentration of charge carriers in the conduction band, obtained through the Hall effect, for various Ga_2O_3 samples doped with Si or Ge donors, obtained by Adam T. Neal, Shin Mou, Subrina Rafique, Hongping Zhao, Elaheh Ahmadi, James S. Speck, Kevin T. Stevens, John D. Blevins, Darren B. Thomson, Neil Moser, Kelson D. Chabak, and Gregg H. Jessen, *Donors and deep acceptors in $\beta-Ga_2O_3$*, App. Phys. Lett. **113**, 062101 (2018). Reproduced with permission of the publisher AIP Publishing LLC.

(b) Assuming that the compensating acceptors are structural vacancies, which structural vacancy can be responsible for the observed behavior? Justify.

(c) A concentration of acceptors 4.8×10^{16} cm^{-3} was obtained for sample 1. Estimate the concentration of donors in this sample.

(d) Estimate the ionization energy of the Si and Ge donors.

(e) Experimental data show that the ionization energy of the donors decreases with increasing donor concentration, for concentrations above 4×10^{17} cm^{-3}. The authors attribute this effect to the formation of an impurity band when the impurity wave functions overlap.

i. Determine the radius of the donors in the effective mass model.

ii. Estimate the concentration of donors above which the donors occupy the entire volume of the sample. Is this result consistent with the conclusion of the authors?

3

Transport properties

3.1 Introduction

We now present some of the basic properties of semiconducting materials resulting from the characteristics presented in the previous chapter and which are core aspects of this class of materials: the existence of a gap of forbidden energies and the huge impact of tiny concentrations of impurities and/or defects on the concentration of charge carriers. The presence of the bandgap strongly affects the optical properties and the concentration of charge carriers due to impurities and defects strongly impacts the transport properties. We therefore limit the analysis of semiconductor properties to these two important categories. In this chapter we will focus on the transport properties.

3.2 Transport phenomena

The term «transport phenomena» applies not only to the transport of charge through the material but also to the transport of heat. As mentioned in Chapter 1, these two phenomena are related through the Wiedemann-Franz law. The transport of electric charge is however of wider significance. We thus focus on it and leave a brief reference to heat transport phenomena at the end of this chapter.

Electric charge transport phenomena can be separated into two subcategories: transport due to the action of an electric field in the material (possibly together with a magnetic field), usually referred to as *drift*; and transport due to the action of a gradient of the concentration of charge carriers, usually referred to as *diffusion*. As we shall see, both these subcategories can, however, be described as a function of a chemical (electro-)potential gradient through the material.

3.2.1 Electrical conduction and scattering mechanisms

The proportionality between the electric current density and the magnitude of the applied electric field (Ohm law), expressed as the electrical conductivity,

DOI: 10.1201/9781003587736-3

implies the existence of dispersion phenomena of electrons as they move through the material. Before addressing the mechanisms responsible for this dispersion, we first consider the dynamics of a conduction electron subject to a uniform electric field $\boldsymbol{\varepsilon}$ only. Although this remains an extremely challenging situation from the experimental point of view, since it implies freezing all dispersion mechanisms, it nevertheless remains very illuminating as a limit situation.

3.2.1.1 Movement of conduction electrons in the absence of scattering: Bloch oscillations

According to eqs. 1.54 and 1.57, the dynamics of a Bloch electron can be described by:

$$\mathbf{F}_{ext} = m^* \frac{d\mathbf{v}}{dt} = \hbar \frac{d\mathbf{k}}{dt} = -e\boldsymbol{\varepsilon}, \tag{3.1}$$

and we immediately get

$$d\mathbf{k} = -\frac{e\boldsymbol{\varepsilon}}{\hbar} dt \rightarrow \mathbf{k}(t) = \mathbf{k}(t_0) - \frac{e\boldsymbol{\varepsilon}}{\hbar} t \tag{3.2}$$

According to eq. 3.2, the (pseudo-)momentum $\mathbf{k}$ varies linearly with time (*ballistic* regime). In the simple approach of the Sommerfeld model, this implies that the entire Fermi sphere corresponding to the occupied states moves linearly with time in k−space, as illustrated in figure 3.1.

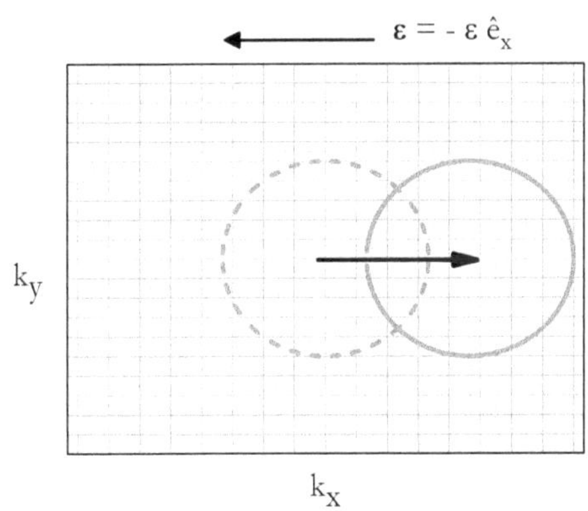

FIGURE 3.1
Displacement of the Fermi sphere in k−space by the application of a uniform electric field $\boldsymbol{\varepsilon} = -\varepsilon\, \hat{e}_x$.

Remember that the bands can often be approximated by a sinusoidal expression of the energy in k−space (eq. 1.83):

$$E(k) = E_0 - 2\gamma \cos{(a\, k)} \tag{3.3}$$

where a is the lattice parameter. The linear dependence of k on time thus translates into a sinusoidal dependence of energy with time (using $\mathbf{k}(t_0) = 0$ without loss of generality):

$$E(k) = E_0 - 2\gamma \cos\left[-\frac{e\,a\,\varepsilon}{\hbar}t\right] \tag{3.4}$$

From eq. 1.51, we obtain the corresponding sinusoidal dependence of the velocity with time. Assuming, for simplicity, a one-dimensional movement due to an electric field $\boldsymbol{\varepsilon} = -\varepsilon\hat{e}_x$, this results in

$$v = \frac{1}{\hbar}\frac{dE}{dk} \rightarrow v(t) = \frac{2\gamma a}{\hbar}\sin(a\,k) = \frac{2\gamma a}{\hbar}\sin\left(\frac{a\,e\,\varepsilon}{\hbar}t\right) \tag{3.5}$$

and the corresponding expression for the position is:

$$x(t) = -\frac{2\gamma}{e\,\varepsilon}\cos\left(\frac{a\,e\,\varepsilon}{\hbar}t\right) \tag{3.6}$$

The result 3.6 is extraordinary: an electron at the bottom of a conduction band (typically described by eq. 3.3), subject to an uniform electric field, oscillates sinusoidally with angular frequency:

$$\omega_B = \frac{a\,e\,\varepsilon}{\hbar} \tag{3.7}$$

These oscillations of conduction electrons subject to a uniform electric field, in the absence of scattering, are called *Bloch oscillations*. For typical values of the parameters ($a \sim 1$ Å, $\varepsilon \sim 1$ V/m), this results in $\omega_B \sim 10^5$ s^{-1}, in the microwave region. The experimental observation of these oscillations requires, however, that the corresponding period $\tau_B = 2\pi/\omega_B$ be much longer than the average scattering time.[1]

3.2.1.2 Free electron gas motion in the presence of scattering

As described in eq. 3.2 and in figure 3.1, the (pseudo-)momentum $\mathbf{k}$ varies linearly with time in the presence of a uniform electric field. In the corresponding Sommerfeld model, this implies that the entire Fermi sphere of the occupied states moves linearly with time in k−space. The displacement of the Fermi sphere over the time interval Δt is thus (assuming for simplicity a one-dimensional description):

$$\Delta k = -\frac{e\,\varepsilon}{\hbar}\Delta t \tag{3.8}$$

Assuming that the conduction electrons are initially occupying the states close to the bottom of the conduction band, we can describe the dispersion

[1] Experimental observation of Bloch oscillations in semiconductor systems has only been realized in semiconductor superlattices (see Chap. 6). See, for example, V. G. Lyssenko *et al.*, *Direct Measurement of the Spatial Displacement of Bloch-Oscillating Electrons in Semiconductor Superlattices*, Physical Review Letters **79**, 301 (1997).

relation in the parabolic approximation:

$$E(k) = \frac{\hbar^2 k^2}{2\,m^*} \tag{3.9}$$

with the respective velocity

$$v = \frac{1}{\hbar}\frac{dE}{dk} = \frac{\hbar}{m^*}k = \frac{\hbar}{m^*}k(t=0) - \frac{e}{m^*}\varepsilon\, t \tag{3.10}$$

The increase of velocity in the time interval Δt is therefore:

$$\Delta v = \frac{\hbar}{m^*}\Delta k = -\frac{e}{m^*}\varepsilon\,\Delta t \tag{3.11}$$

In the presence of scattering mechanisms and in the approximation of the relaxation time described in Chapter 1 (Drude model), the dispersion mechanism is described by an average time τ. We consider that this interaction is strong enough so that, on average, the velocity of the conduction electrons after the interaction is zero, and the electrons return to the levels close to the bottom of the band.

We also consider that the relaxation time is short enough for the electrons to remain close to the bottom of the band, so that the parabolic approximation remains valid. In this description, conduction electrons moving in the course of an average τ time between collisions achieve an average speed:

$$< v >= -\frac{e}{m^*}\tau\,\varepsilon = -\mu_e\varepsilon \tag{3.12}$$

This defines the *mobility* μ_e of the conduction electrons:

$$\mu_e = \frac{e\,\tau}{m^*}, \tag{3.13}$$

which describes the proportionality between the drift velocity of conduction electrons and the applied electric field.[2] Note also that eq. 3.12 can be interpreted as corresponding to the condition of equilibrium in the presence of an electric field and viscous friction of the form $F_a = -p/\tau$:

$$F_a + F_e = 0 \Leftrightarrow -\frac{p}{\tau} - e\varepsilon = 0 \tag{3.14}$$

The average density of current corresponding to eq. 3.12 is:

$$< J >= -n\,e < v >= -n\,e(-\mu_e)\varepsilon = \frac{n\,e^2\tau}{m^*}\varepsilon \tag{3.15}$$

and the corresponding conductivity is:

$$\sigma = \frac{ne^2\tau}{m^*} \tag{3.16}$$

[2]There is an ambiguity in the literature, a source of perennial confusion, regarding the sign of conduction electron mobility. Equation 3.12 invites us to define the mobility of conduction electrons as a negative quantity, thus expressing the fact that the electrons move in the opposite direction to the applied field. However, the values found in the literature generally do not include the sign. We have therefore adopted the corresponding definition of conduction electron mobility as a positive quantity.

Expression 3.16 illustrates the complexity of the physical phenomena associated with conductivity, which therefore depends on dispersion phenomena (expressed in the average time τ), the band structure (expressed in the effective mass m^*) and the concentration of charge carriers n. Of these parameters, the concentration of charge carriers is controllable through the concentration of impurities, as we saw in Chapter 2. The average time τ and the effective mass m^*, as well as the corresponding mobility μ_e, are generally much more difficult to control for a given material.

In the presence of electrons and holes, the total current is simply the sum of the respective currents:

$$\mathbf{J} = \mathbf{J_e} + \mathbf{J_h} = -en\mathbf{v_e} + ep\mathbf{v_h} \tag{3.17}$$

Taking into account that $\mathbf{v_e} = -\mu_e\mathbf{E}$ and that $\mathbf{v_h} = \mu_h\mathbf{E}$ (with $\mu_e = e\tau_e/m_e^*$ and
$\mu_h = e\tau_h/m_h^*$, respectively):

$$\mathbf{J} = (en\mu_e + ep\mu_h)\,\mathbf{E} \tag{3.18}$$

and the total conductivity becomes:

$$\sigma = en\mu_e + ep\mu_h \tag{3.19}$$

3.2.2 Sources of dispersion

According to Bloch theorem, electrons in perfect crystals are described by a modulated plane wave and are not subject to dispersion mechanisms. Bloch theorem does not therefore contain Ohm law, which is the result of the existence of deviations (impossible to avoid) from perfect periodicity.

In this sense, any deviation from the crystal's periodicity constitutes a possible dispersion mechanism. Among the most important mechanisms in *bulk* (the massive interior of the crystal, where the effects due to the break in periodicity on the surface are negligible) are those due to impurities/defects and those due to the atomic movements around the equilibrium position in the lattice, caused by vibrations (phonons). The diversity of possible dispersion mechanisms in *bulk* is thus immense. However, in covalent crystals (Si, Ge), dispersion by ionized impurities and dispersion due to the deformation potential are particularly important.

3.2.2.1 Dispersion by ionized impurities

The scattering of charge carriers is a mechanism similar to Rutherford[3] scattering. The treatment of this process goes beyond the scope of this book: it requires the consideration of a shielded potential and leads to a mobility that increases with temperature according to $T^{3/2}$.

[3]Ernest Rutherford (1871-1937).

3.2.2.2 Deformation potential

The dispersion of charge carriers by the lower energy vibrations corresponding to long-wavelength acoustic phonons is usually called *deformation potential.* The corresponding deformation can have a transverse character, corresponding to torsion waves, or a longitudinal character, corresponding to compression waves. The deformation potential leads to a mobility that decreases with temperature according to $T^{-3/2}$.

3.2.2.3 Effective mobility

In the presence of several dispersive processes with characteristic times τ_1, τ_2, etc., each will contribute a term of the form $-\mathbf{v}/\tau_i$ to the acceleration:

$$\frac{d\mathbf{v}}{dt} = -\frac{\mathbf{v}}{\tau_1} - \frac{\mathbf{v}}{\tau_2} - \frac{\mathbf{v}}{\tau_3} - \ldots = -\left(\Sigma_i \frac{1}{\tau_i}\right)\mathbf{v} = -\frac{\mathbf{v}}{\tau} \tag{3.20}$$

where the effective time τ is:

$$\frac{1}{\tau} = \Sigma_i \frac{1}{\tau_i}. \tag{3.21}$$

Given that the mobility is $\mu = e\,\tau/m^*$, the effective mobility as a function of the mobility of each process also results in:

$$\frac{1}{\mu} = \Sigma_i \frac{1}{\mu_i}. \tag{3.22}$$

It should be noted that in the presence of two processes with very different mobilities, the effective mobility is dominated by the process with the *lower* mobility. Thus, in covalent crystals, dispersion by ionized impurities dominates at low temperatures and the deformation potential dominates at high temperatures.

It should also be noted that mobilities typically decrease with increasing impurity concentration, so high concentrations and high mobilities are mutually exclusive.

3.2.3 Conduction in the presence of magnetic fields

We now analyze the dynamics of conduction electrons subject to scattering in the simultaneous presence of an applied electric field and a magnetic field, as shown in figure 3.2. This is the basic device for observing the Hall effect, which is routinely used to characterize the concentration and types of charge carriers in semiconductors.

In addition to the Lorentz[4] force $\mathbf{F_L} = -e(\mathbf{E}+\mathbf{v}\times\mathbf{B})$ we must also include the effect of dispersion translated into viscous friction of the type $\mathbf{F_a} = -\mathbf{p}/\tau$. Eq. 1.54 thus results:

$$\hbar\frac{d\mathbf{k}}{dt} = -e(\mathbf{E}+\mathbf{v}\times\mathbf{B}) - \frac{\mathbf{p}}{\tau} \tag{3.23}$$

[4]Hendrik Antoon Lorentz (1853-1928).

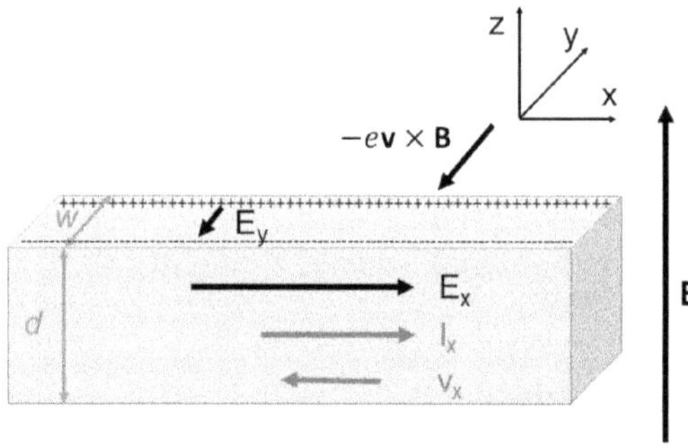

FIGURE 3.2
Basic diagram of the Hall effect: a current $\mathbf{j} = j_x\,\hat{\mathbf{e}}_\mathbf{x}$ flows due to an applied electric field $\mathbf{E_a} = E_x\,\hat{\mathbf{e}}_\mathbf{x}$, in the presence of a magnetic field $\mathbf{B} = B_0\,\hat{\mathbf{e}}_\mathbf{z}$. The situation of negative charge carriers is represented, where the Lorenz force and the Hall electric field point in the negative direction of the y-axis (in the reference frame of the figure). The figure also shows the thickness d of the sample, as well as its width w.

Once again, we consider that the average scattering time τ is sufficiently short and the fields are sufficiently weak for the electrons to remain close enough to the bottom of the band so that the parabolic approximation remains valid. We can therefore apply eq. 3.11, which allows us to rewrite eq. 3.23 in the form (assuming $v(t=0)=0$):

$$m^* \left[\frac{d}{dt} + \frac{1}{\tau}\right] \mathbf{v} = -e(\mathbf{E} + \mathbf{v} \times \mathbf{B}) \tag{3.24}$$

Considering the basic geometry of figure 3.2, where $\mathbf{B} = B_0\hat{\mathbf{e}}_\mathbf{z}$, it follows:

$$\begin{aligned} m^* \left[\frac{d}{dt} + \frac{1}{\tau}\right] v_x &= -e\,E_x - e\,v_y\,B_0 \\ m^* \left[\frac{d}{dt} + \frac{1}{\tau}\right] v_y &= -e\,E_y + e\,v_x\,B_0 \\ m^* \left[\frac{d}{dt} + \frac{1}{\tau}\right] v_z &= -e\,E_z \end{aligned} \tag{3.25}$$

In the steady-state, where $d/dt \longrightarrow 0$:

$$\begin{aligned} v_x &= -\mu_e\,E_x - \omega_c\tau\,v_y \\ v_y &= -\mu_e\,E_y + \omega_c\tau\,v_x \\ v_z &= -\mu_e\,E_z \end{aligned} \tag{3.26}$$

where the expression for mobility (eq. 3.13) and the classical expression for the cyclotron resonance frequency were used.[5] It is possible to rewrite eq. 3.26 in order to highlight the linear relation between velocity and the total electric field. Given $\mathbf{J} = -n\,e\,\mathbf{v}$, Ohm law in the presence of a magnetic field can be written in the form $\rho\mathbf{J} = \mathbf{E}$, where ρ is now the magnetoresistivity tensor:

$$\rho = \frac{1}{\sigma_0}\begin{pmatrix} 1 & \omega_c\tau & 0 \\ -\omega_c\tau & 1 & 0 \\ 0 & 0 & 1 \end{pmatrix} \tag{3.27}$$

$\sigma_0 = n\,e^2\,\tau/m^*$ is the static conductivity in the absence of magnetic field. In the steady state and in the geometry of the figure 3.2, the current is only flowing in the X direction: $\mathbf{J} = J_x\,\hat{e}_x$ and Ohm law in the tensorial form $\rho\mathbf{J} = \mathbf{E}$ becomes:

$$\frac{1}{\sigma_0}\begin{pmatrix} 1 & \omega_c\tau & 0 \\ -\omega_c\tau & 1 & 0 \\ 0 & 0 & 1 \end{pmatrix}\begin{pmatrix} J_x \\ 0 \\ 0 \end{pmatrix} = \begin{pmatrix} E_x \\ E_y \\ E_z \end{pmatrix} \tag{3.28}$$

We therefore obtain the usual relation $J_x = \sigma_0\,E_x = (n\,e^2\,\tau/m^*)E_x$. We also get, for the y equation, the value of the steady-state electric field in this direction (the Hall field E_H):

$$E_H = E_y = \rho_{yx}J_x = -\omega_c\tau\,E_x = -\mu_e\,B_0\,E_x = -\frac{\mu_e}{\sigma_0}\,B_0\,J_x = -\frac{1}{n\,e}\,B_0\,J_x \tag{3.29}$$

Note that the Hall field (and the corresponding Hall voltage $V_H = E_y\,w$) grows trivially with the intensity of the applied fields B_0 and E_x (and therefore with the current J_x). It is therefore usual to define the Hall coefficient R_H as the ratio $R_H = E_y/(J_x\,B_0)$. In the simplest case where the conductivity is due to conduction electrons only, we get:

$$R_H = \frac{E_y}{J_x\,B_0} = \frac{\rho_{yx}}{B} = -\frac{1}{n\,e} \tag{3.30}$$

The Hall coefficient is thus negative and identical to the inverse of the charge concentration. The *Hall conductivity*, σ_H (and the corresponding resistivity $\rho_H = 1/\sigma_H$) is also usually defined from the relation:

$$J_x = \sigma_H\,E_y \longrightarrow \rho_H = \frac{E_y}{J_x} = B\,R_H \tag{3.31}$$

We can also rewrite eq. 3.26 in the tensorial form $\mathbf{J} = \sigma\mathbf{E}$ as follows:

$$\begin{pmatrix} v_x \\ v_y \\ v_z \end{pmatrix} = \begin{pmatrix} -\frac{\mu_e}{1+(\omega_c\tau)^2} & \frac{\mu_e\omega_c\tau}{1+(\omega_c\tau)^2} & 0 \\ -\frac{\mu_e\omega_c\tau}{1+(\omega_c\tau)^2} & -\frac{\mu_e}{1+(\omega_c\tau)^2} & 0 \\ 0 & 0 & \frac{\omega_c\tau}{B} \end{pmatrix}\begin{pmatrix} E_x \\ E_y \\ E_z \end{pmatrix} \tag{3.32}$$

[5] Recall that a classical particle with charge e and velocity v, subject to a magnetic field B perpendicular to its velocity, performs a uniform circular motion with centripetal acceleration $v^2/r = e\,v\,B/m$ and velocity $v = 2\pi\,r/T = \omega_c\,r$, so the angular frequency of the motion is $\omega_c = e\,B/m$.

Again, given $\mathbf{J} = -n\,e\,\mathbf{v}$, we now obtain Ohm law in the presence of a magnetic field $\mathbf{J} = \sigma\mathbf{E}$, where σ is now the *static magneto-conductivity tensor*:

$$\sigma = \frac{\sigma_0}{1+(\omega_c\tau)^2}\begin{pmatrix} 1 & -\omega_c\tau & 0 \\ \omega_c\tau & 1 & 0 \\ 0 & 0 & 1+(\omega_c\tau)^2 \end{pmatrix} \tag{3.33}$$

3.2.3.1 Hall effect in the presence of electrons and holes

As mentioned in the previous chapter, the charge carriers in semiconductors can be either conduction electrons or holes, and the case where both types of charge carriers are present simultaneously should be considered. The sum of the currents corresponding to each of the carriers becomes:

$$\mathbf{J} = \mathbf{J_e} + \mathbf{J_h} = \sigma_e\,\mathbf{E} + \sigma_h\,\mathbf{E} \tag{3.34}$$

The respective magnetoconductivity tensors must then be added. As before, in the stationary case and in the geometry of figure 3.2, we have

$$J_y = 0 \Rightarrow \sigma_{yx}E_x + \sigma_{yy}E_y = 0 \Rightarrow E_y = -\frac{\sigma_{yx}}{\sigma_{yy}}E_x \tag{3.35}$$

The J_x current is therefore:

$$J_x = \sigma_{xx}E_x + \sigma_{xy}E_y = \frac{\sigma_{xx}\sigma_{yy} - \sigma_{xy}\sigma_{yx}}{\sigma_{yy}}E_x, \tag{3.36}$$

and the corresponding Hall coefficient is (see Problem 5):

$$R_H = \frac{1}{e}\frac{-n\mu_e^2(1+\mu_h^2B^2) + p\mu_h^2(1+\mu_e^2B^2)}{n^2\mu_e^2(1+\mu_h^2B^2) - 2np\mu_e\mu_h(1+\mu_e\mu_hB^2) + p^2\mu_h^2(1+\mu_e^2B^2)} \tag{3.37}$$

In the low-field limit ($\mu^2B^2 << 1$) this Hall coefficient becomes

$$R_H = \frac{1}{e}\frac{p\,\mu_h^2 - n\,\mu_e^2}{(p\,\mu_h - n\,\mu_e)^2} \tag{3.38}$$

This expression reduces to $1/pe$ or $-1/ne$ if the transport is dominated by holes or electrons, respectively. The Hall effect at low fields thus provides an immediate indication of the type of majority charge carriers. However, the experiment must be carried out at relatively low temperatures: in fact, as the temperature increases and bearing in mind that in general $\mu_e > \mu_h$, it can happen that $n\,\mu_e^2 > p\,\mu_h^2$, leading to a reversal of the sign of the Hall coefficient.

3.2.4 Conduction in high electric fields

Up to now we have assumed that the applied electric field is sufficiently low for the kinetic energy acquired by the electrons to remain of the order of the thermal energy ($3\,k\,T/2$) or lower, so the relaxation time approximation we have been using is valid. This approximation implies that dispersion occurs randomly and independently of the velocity of the electron. Only its dependence with temperature plays a significant role.

The relaxation time approximation is broken when the electric field is sufficiently intense for the electrons to gain kinetic energy far in excess of the thermal energy (*hot electrons*). The increase in velocity of these electrons is now limited by the probability of energy transfer to the lattice (or to the electronic fluid) through the creation of excitations (particularly vibrations). The drift velocity thus no longer follows the linear dependence with the applied field typical of ohmic behavior, and tends to saturate at a limit value. One of the most important mechanisms is the emission of a longitudinal optical phonon; the threshold velocity value is reached, *grosso modo*, when the kinetic energy of the electron becomes similar to the phonon's excitation energy, $\hbar\omega_{LO}$:

$$\frac{1}{2}m^* v_s^2 \sim \hbar\omega_{LO} \Rightarrow v_s \sim \sqrt{\frac{2\hbar\omega_{LO}}{m^*}} \tag{3.39}$$

3.2.5 Negative differential resistivity

Figure 3.3 shows experimental and theoretical data for the drift velocity in GaAs as a function of the applied electric field. The figure clearly illustrates the linear ohmic dependence at low fields, which ceases for fields of the order of 3–4 kV/cm. However, for higher fields, there is a marked decrease in drift velocity as the applied electric field increases, an unexpected phenomenon known as *negative differential resistivity*.

The explanation for this phenomenon lies in the band structure of GaAs (figure 3.4), characterized by two conduction band minima at the Γ and L points with different effective masses m^*_Γ and m^*_L, respectively, where $m^*_\Gamma < m^*_L$. For low fields, electrons only occupy levels around the absolute minimum at the Γ point and are characterized by the mobility $\mu_\Gamma = e\tau/m^*_\Gamma$. However, as the electric field increases and the electrons gain kinetic energy exceeding thermal energy, some of the electrons reach the metastable minimum at the point L, which is gradually populated. The mobility of these electrons is $\mu_L = e\tau/m^*_L < \mu_\Gamma$, lower than the mobility at the Γ point. Thus, for progressively higher fields, the population of electrons with reduced mobility grows, which leads to a progressive decrease in average mobility and a decrease in drift velocity. The value of the electric field for which the drift velocity begins to decrease is called *critical field.*

A consequence of negative differential resistivity and of populating the L point is the occurrence of a spatial instability in the charge distribution (which is similar to, but not to be confused with the oscillations in real space discussed at the beginning of the chapter – Bloch oscillations). For fields close to the critical field, electric field fluctuations can lead to a spatial zone – called *domain* – where the electric field is higher than the critical field, and the speed of the conduction electrons is lower than outside. Thus, in the

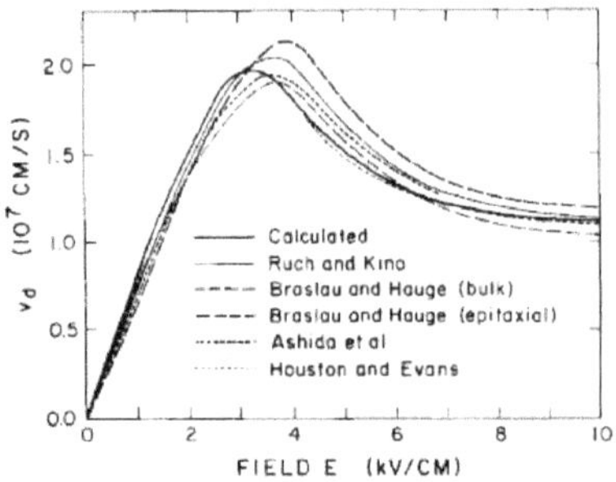

FIGURE 3.3
Drift velocity in GaAs as a function of applied electric field: experimental and theoretical values (J. S. Blakemore, *Semiconducting and other major properties of gallium arsenide*, J. App. Phys. **53**, R123 (1982), Fig. 57, and references therein). Reproduced with permission of the publisher AIP Publishing LLC.

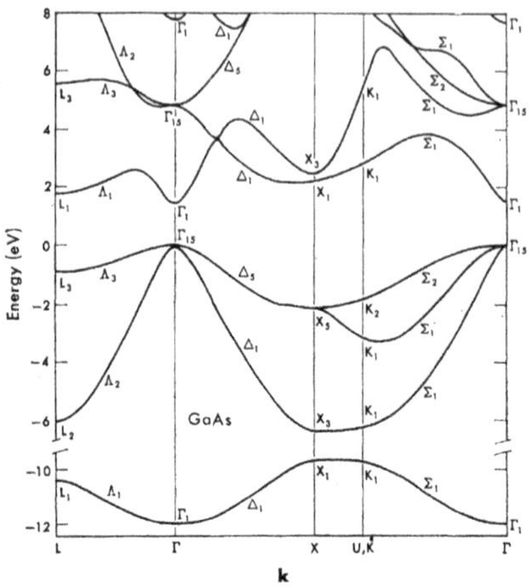

FIGURE 3.4
Band structure of GaAs, calculated in the directions of greatest symmetry of the Brillouin zone. (Figure reproduced with permission from J. P. Walter and M. L. Cohen, *Calculation of the Reflectivity, Modulated Reflectivity, and Band Structure of GaAs, GaP, ZnSe, and ZnS*, Phys. Rev. **183**, 763 (1969). *Copyright* (1969) of the American Physical Society).

outer zones bordering the domain, the conduction electrons will have higher velocities, leading to the accumulation of charge on one side to the depletion of charge on the other side.[6] This instability gives rise to a time-dependent current corresponding to the displacement of the domain, which allows GaAs to be used as a microwave emitter (*Gunn effect*), forming the device called *Gunn diode*.[7]

3.2.6 Diffusion

As mentioned in the introduction to this chapter, charge transport in materials can also occur due to gradients in the concentration of charge carriers, which

[6]Cf. Peter Y. Yu and Manuel Cardona, *Fundamentals of Semiconductors: Physics and Materials Properties*, 2nd edition, pp. 220-222.

[7]John Gunn (1928-2008). The designation "diode" is inappropriate, as this is not a device that only allows current to flow in one direction, as we will see in Chapter 5.

are the cause of particle currents in the direction of lower concentrations. To a first approximation, the particle current is simply proportional to the gradient of concentration and flows in the opposite direction (*Fick law*[8]). The density of electric current associated to density of particles with charge q, due to a concentration gradient ϕ is therefore:

$$\mathbf{J} = -q\, D\, \nabla \phi, \tag{3.40}$$

where D is a proportionality constant designated *diffusion coefficient*. For electrons and holes in semiconductors, the respective current densities $\mathbf{J_n}$ and $\mathbf{J_p}$, associated with gradients in the respective concentrations are written as:

$$\mathbf{J_n} = e\, D_n\, \nabla n \tag{3.41}$$
$$\mathbf{J_p} = -e\, D_p\, \nabla p, \tag{3.42}$$

where D_n and D_p are the diffusion coefficients of conduction electrons and holes, respectively. When a concentration gradient and an electric field are simultaneously present, the respective current densities are as follows:

$$\begin{aligned} \mathbf{J_n} &= e\, \mu_n\, n\mathbf{E} + e\, D_n\, \nabla n \\ \mathbf{J_p} &= e\, \mu_p\, p\mathbf{E} - e\, D_p\, \nabla p, \end{aligned} \tag{3.43}$$

3.2.6.1 Einstein relations

Both in the case of drift and of diffusion, the movement of the same charges is considered. It is therefore not entirely surprising that a relation can be established between the mobility and the diffusion coefficient. In order to obtain that relation, we start with the general expressions for the concentration n of conduction electrons and the concentration p of holes (eqs. 2.10 and 2.11):

$$\begin{aligned} n &= N_C \exp\left(\frac{\mu - E_C}{kT}\right) \\ p &= N_V \exp\left(\frac{E_V - \mu}{kT}\right) \end{aligned} \tag{3.44}$$

In the presence of an electric potential V, the conduction electrons are subject to an additional potential energy $-e\,V$. The bottom of the conduction band therefore corresponds to the energy $E_C - e\,V$ and the top of the valence band corresponds to the energy $E_V - e\,V$. The expressions for the concentrations are correspondingly changed to:

$$\begin{aligned} n &= N_C \exp\left(\frac{\mu_{eq} - (E_C - e\,V)}{kT}\right) = N_C \exp\left(\frac{(\mu_{eq} + e\,V) - E_C}{kT}\right) \\ p &= N_V \exp\left(\frac{(E_V - e\,V) - \mu_{eq}}{kT}\right) = N_V \exp\left(\frac{E_V - (\mu_{eq} + e\,V)}{kT}\right) \end{aligned} \tag{3.45}$$

[8] Adolf Fick (1829-1901).

Note that the quantity μ_{eq} corresponding to the variation of free energy with the number of particles now has two contributions: an external contribution due to the electrostatic potential energy per electron $(-eV)$ and an «internal» contribution, identical to the chemical potential μ_q in the absence of an electric field:

$$\mu_{eq} = \mu_q - eV \tag{3.46}$$

The quantity μ_{eq} is thus often referred to as the *electrochemical potential*,[9] and results explicitly, from eq. 3.45:

$$\begin{aligned} \mu_{eq} &= -eV + E_C + kT \ln \frac{n}{N_C} \\ \mu_{eq} &= -eV + E_V - kT \ln \frac{p}{N_V} \end{aligned} \tag{3.47}$$

In the presence of an electric field, the internal energy of the system is now dependent not only on the internal degrees of freedom (associated with the chemical potential) but also on the potential energy of each particle in the presence of the electrostatic potential. The condition of uniformity of the chemical potential characteristic of thermodynamic equilibrium is thus now replaced by the condition of uniformity of the electrochemical potential.

The relation between the gradients in electrochemical potential, electrostatic potential and concentration becomes (using $\mathbf{E} = -\nabla V$):

$$\begin{aligned} \nabla \mu_{eq} &= e\mathbf{E} + \frac{kT}{n} \nabla n \\ \nabla \mu_{eq} &= e\mathbf{E} - \frac{kT}{p} \nabla p \end{aligned} \tag{3.48}$$

Equations 3.43 and 3.48 express the same reality, so the ratio between the terms proportional to $\mathbf{E}$ and ∇n or ∇p must be the same in both equations. We therefore obtain:

$$\begin{aligned} \mu_n &= \frac{e\,D_n}{k\,T} \\ \mu_p &= \frac{e\,D_n}{k\,T} \end{aligned} \tag{3.49}$$

Mobilities and diffusion constants are thus related. These results are known as *Einstein relations*. These relations allow us to rewrite eq. 3.43. Taking eqs. 2.10 and 2.11 into account, we begin by calculating the concentration

[9] Cf. John Bardeen, *Surface States and Rectification at a Metal Semi-Conductor Contact*, Physical Review **71**, 717 (1947), for a particularly clear presentation.

gradients:

$$\begin{aligned}\nabla n &= \frac{N_C}{kT}\exp\left(\frac{\mu_q - E_C}{kT}\right)\nabla\mu_q = \frac{1}{kT}n\nabla\mu_q \\ \nabla p &= \frac{N_V}{kT}\exp\left(\frac{E_V - \mu_q}{kT}\right)\nabla\mu_q = -\frac{1}{kT}p\nabla\mu_q\end{aligned} \tag{3.50}$$

Now, taking Einstein relations into account, eq. 3.43 becomes:

$$\mathbf{J_n} = e\,\mu_n\, n\left(\mathbf{E} + \frac{1}{e}\nabla\mu_q\right) = e\,\mu_n\, n\mathbf{E_{eff}} \tag{3.51}$$

$$\mathbf{J_p} = e\,\mu_p\, p\left(\mathbf{E} + \frac{1}{e}\nabla\mu_q\right) = e\,\mu_p\, p\mathbf{E_{eff}} \tag{3.52}$$

where we have defined the "effective field" $\mathbf{E_{eff}}$ describing both drift and diffusion as

$$\mathbf{E_{eff}} = \mathbf{E} + \frac{1}{e}\nabla\mu_q \tag{3.53}$$

3.2.7 Continuity equations

The law of conservation of electric charge is expressed in local form through the well-known continuity equation that relates the electric charge density ρ to the current density $\mathbf{J}$:

$$\frac{\partial\rho}{\partial t} = -\mathrm{div}\mathbf{J} \tag{3.54}$$

In the absence of the creation or recombination of electron-hole pairs, it is possible to write continuity equations for electrons in the conduction band and for holes in the valence band

$$\begin{aligned}\frac{\partial(-e\,n)}{\partial t} &= -\mathrm{div}\mathbf{J_n} \\ \frac{\partial(e\,p)}{\partial t} &= -\mathrm{div}\mathbf{J_p}\end{aligned} \tag{3.55}$$

The rates of change in the concentrations of conduction electrons and holes are therefore:

$$\begin{aligned}\frac{\partial n}{\partial t} &= \frac{1}{e}\mathrm{div}\mathbf{J_n} \\ \frac{\partial p}{\partial t} &= -\frac{1}{e}\mathrm{div}\mathbf{J_p}\end{aligned} \tag{3.56}$$

If conduction electrons and holes are created at rates G_n and C_p, respectively, and conduction electrons and holes are recombined at rates U_n and U_p, respectively, these continuity equations can be rewritten as:

$$\begin{aligned}\frac{\partial n}{\partial t} &= G_n - U_n + \frac{1}{e}\mathrm{div}\mathbf{J_n} \\ \frac{\partial p}{\partial t} &= G_p - U_p - \frac{1}{e}\mathrm{div}\mathbf{J_p}\end{aligned} \tag{3.57}$$

Taking eq. 3.43 into account, this results in:[10]

$$\begin{aligned}\frac{\partial n}{\partial t} &= G_n - U_n + \mu_n\, n\,\mathrm{div}\mathbf{E} + \mu_n\,\mathbf{E}\cdot\nabla n + D_n\,\nabla^2 n \\ \frac{\partial p}{\partial t} &= G_p - U_p - \mu_p\, p\,\mathrm{div}\mathbf{E} + \mu_p\,\mathbf{E}\cdot\nabla p + D_p\,\nabla^2 p\end{aligned} \tag{3.58}$$

In the absence of an electric field, these equations become:

$$\begin{aligned}\frac{\partial n}{\partial t} &= G_n - U_n + D_n\,\nabla^2 n \\ \frac{\partial p}{\partial t} &= G_p - U_p + D_p\,\nabla^2 p\end{aligned} \tag{3.59}$$

In the steady-state $\partial n/\partial t = 0 = \partial p/\partial t$, we get the important result:

$$\begin{aligned}\nabla^2 n &= -\frac{G_n - U_n}{D_n} \\ \nabla^2 p &= -\frac{G_p - U_p}{D_p}\end{aligned} \tag{3.60}$$

3.2.8 Thermoelectric properties

Heat transport is closely linked to charge transport, as mentioned in Chapter 1 and as expressed by the Wiedemann-Franz law. This connection is also apparent in some electrical effects caused by temperature gradients, which we briefly mention in this section.

We start by considering that the transport of heat due to a temperature gradient ∇T is described by Fourier law:

$$\mathbf{J_Q} = -\kappa\,\nabla T \tag{3.61}$$

where $\mathbf{J_Q}$ is the energy per unit area and time that flows in the direction of the temperature gradient (and in the direction of decreasing temperatures); κ is the proportionality coefficient designated *thermal conductivity*. This «current» of energy $\mathbf{J_Q}$ obeys a continuity equation:

$$\mathrm{div}\mathbf{J_Q} = -\frac{\partial Q}{\partial t}, \tag{3.62}$$

[10] We used the well-known results of differential calculus $\mathrm{div}(n\,\mathbf{E}) = n\,\mathrm{div}\mathbf{E} + \mathbf{E}\cdot\nabla n$ and $\mathrm{div}(\nabla n) = \nabla^2 n$.

where Q represents the thermal energy (heat) per unit volume, and:

$$Q = \rho C T, \tag{3.63}$$

ρ is the density of the material and C is its heat capacity. From the continuity equation (eq. 3.62), we obtain:

$$\mathrm{div}\mathbf{J_Q} = -\rho C \frac{\partial T}{\partial t}, \tag{3.64}$$

If heat is generated at rate A (due to a heat source), the corresponding term must be added to eq. 3.64:

$$\mathrm{div}\mathbf{J_Q} = -\rho C \frac{\partial T}{\partial t} + A, \tag{3.65}$$

Using Fourier law (eq. 3.61), we obtain:

$$\nabla^2 T = \frac{\rho C}{\kappa}\frac{\partial T}{\partial t} - \frac{A}{\kappa} \tag{3.66}$$

In the stationary case and in the absence of heat sources, this equation reduces to the Laplace[11] equation:

$$\nabla^2 T = 0 \tag{3.67}$$

3.2.8.1 Coupled transport of charge and heat

The transport of charge and heat takes place simultaneously as charges flow carrying kinetic energy. The physics of these processes can be analyzed using the tools of statistical physics. A detailed analysis is beyond the scope of this book, so we limit ourselves here to a few illustrative notes as a mere introduction to this topic.

The coupled transport of charge and heat can be described using the equations:

$$\begin{aligned} \mathbf{J} &= \sigma \mathbf{E_{eff}} + L\nabla T \\ \mathbf{J_Q} &= M\mathbf{E_{eff}} + N\nabla T, \end{aligned} \tag{3.68}$$

where $\mathbf{E_{eff}}$ is the effective field defined in eq. 3.53 and which includes both drift and diffusion. σ, L, M and N are proportionality coefficients. The experimentally accessible quantities are the electric current density $\mathbf{J}$ and the temperature gradient ∇T, so it is convenient to rewrite eqs. 3.68 as:

$$\begin{aligned} \mathbf{E_{eff}} &= \rho\mathbf{J} + S\nabla T \\ \mathbf{J_Q} &= \Pi\mathbf{J} - K\nabla T, \end{aligned} \tag{3.69}$$

where $\rho = 1/\sigma$ is the specific resistance, $S = -L/\sigma$ is the thermoelectric power, $\Pi = M/\sigma$ is the Peltier coefficient (thermal energy transported per unit load), and $K = (M\,L/\sigma) - N$.

[11] Pierre-Simon de Laplace (1749-1827).

3.2.8.2 Seebeck effect and Peltier effect

Consider a semiconductor in an open circuit, subject to a temperature gradient ∇T. In this case, the current through the material is zero $\mathbf{J} = 0$, so:

$$\mathbf{E_{eff}} = S \nabla T \tag{3.70}$$

An electric field appears due to a temperature gradient. This effect is called *Seebeck effect*[12] or *thermoelectric effect.* An important aspect is that electrons and holes have different signs of the thermoelectric coefficient, so the Seebeck effect makes it possible to determine the type of carriers.

If an electric current is allowed to pass through, heat transmission occurs at the same time (*Peltier effect*[13]), whose coefficient also depends on the type of carriers.

Problems

1. Superconductivity is a phenomenon that occurs at low temperatures, when electron correlation mechanisms become important, allowing the formation of electron pairs (known as *Cooper pairs*[14]) with spin 1, which therefore behave like bosons and do not obey the Pauli exclusion principle. These pairs move without suffering scattering effects, resulting in the disappearance of the resistivity of the material. When discussing this aspect, comparisons are often made with an imaginary *perfect conductor*, where the electrons remain independent but the dispersion mechanisms are minimized, thus hypothetically leading to zero resistivity.

   (a) Show that, in the case of electrons at the bottom of the conduction band of a one-dimensional crystal ($E(k) = E_0 \sin(a\,k)$), subject to an electric field $\boldsymbol{\varepsilon} = -\varepsilon \hat{e}_x$, the conductivity is:

   $$\sigma = \frac{n\,E_0\,a}{\hbar\varepsilon} \cos\left(\frac{a\,e\,\varepsilon}{\hbar} t\right)$$

   where n is the concentration of electrons in the conduction band.

   (b) Determine the average value of the conductivity. Comment this value.

   (c) Can periodical crystals be «perfect conductors»? What about other non-periodic systems in condensed matter?

[12] August Seebeck (1805-1849).
[13] Jean Peltier (1785-1845).
[14] Leon Cooper (1930-2024).

2. A sample of pure Si has been doped with a concentration of As atoms equal to 10^{18} cm^{-3}. Taking into account that electron and hole mobilities in Si at 300 K are $\mu_e = 1350$ cm^2 V^{-1} s^{-1} and $\mu_h = 475$ cm^2 V^{-1} s^{-1}, respectively, determine the conductivity of the Si sample at this temperature.

3. A sample of Ge is doped with only one type of impurity. If the concentration of donors is 10^{14} cm^{-3}, what are the experimental conditions necessary to observe cyclotron resonance? ($m_e^* = m_0$; the orbit diameter of the donor electron is approximately 300 Å).

4. Obtain eq. 3.33 for the magnetoconductivity tensor, from eq. 3.26.

5. Consider a semiconductor where both types of charge carriers (electrons and holes) are present simultaneously.

   (a) Obtain the following expression for the Hall coefficient:

$$R_H = \frac{1}{e}\frac{-n\mu_e^2(1+\mu_h^2B^2)+p\mu_h^2(1+\mu_e^2B^2)}{n^2\mu_e^2(1+\mu_h^2B^2)-2np\mu_e\mu_h(1+\mu_e\mu_hB^2)+p^2\mu_h^2(1+\mu_e^2B^2)}$$

   (b) Obtain the corresponding limits in the high-field approximation and in the low-field approximation.

6. *Transport at high frequencies.* Consider a material with static conductivity (DC) σ_0. Show that the AC conductivity $\sigma(\omega)$ for the frequency ω is:

$$\sigma(\omega) = \frac{ne^2\tau}{m^*}\frac{1}{1+\omega^2\tau^2} + i\frac{ne^2\tau}{m^*}\frac{\omega\tau}{1+\omega^2\tau^2}$$

7. Indium nitride, InN, is a III–V family semiconductor with a particularly small bandgap ($E_g \approx 0.7$ eV) and with potential applications in fast electronics. The bonds are partially ionic, with the indium atom giving up electrons to the nitrogen atom. This semiconductor tends to exhibit n-type behavior, even without the deliberate introduction of any dopants. The electron drift rate in InN, as a function of the applied electric field, can be described using a model where the conduction band has two «valleys» separated by 1.775 eV, of effective masses 0.04 m_0 and 0.25 m_0. Explain how this model leads to a reduction in the drift velocity as the electric field increases.

8. Cadmium sulphide (CdS) is a II–VI semiconductor with E_g = 2.5 eV. Indium (In) is potentially an excellent donor candidate in CdS, but the electrical activation of this impurity is strongly dependent on the heat treatment conditions of the material. U. V. Desnica *et al.* [15] investigated the temperature dependence of the concentration of free charge carriers and the (Hall) mobility for various CdS

[15] U. V. Desnica, I. D. Desnica-Frankovic, R. Magerle, and M. Deicher, *Compensating defects and electrical activation of donors in CdS*, Physica B **273-274**, 907 (1999).

samples implanted with In and subsequently annealed at different T_a temperatures in a Cd or S atmosphere. Annealing in a S atmosphere strongly suppresses the formation of S atomic vacancies (favoring the formation of Cd atomic vacancies) and annealing in a Cd atmosphere strongly suppresses the formation of Cd atomic vacancies (favoring the formation of S atomic vacancies). Experiments show that the transport properties of CdS do not vary appreciably with annealing in a Cd atmosphere. The experimental data can be adjusted to a condition of electrical neutrality by assuming a donor with a binding energy of 27 meV and partial compensation by an acceptor whose concentration depends on annealing.

$(m_e = 0.16\, m_0,\ m_h = 0.7\, m_0,\ \epsilon_r = 9.12)$

(a) Is the experimentally measured binding energy consistent with CdS donors described by the effective mass model?

(b) Estimate the temperature above which the intrinsic process becomes dominant.

(c) Experimental data show that the concentration of conduction electrons saturates at temperatures above 200 K. The value of the saturation concentration decreases with annealing temperature (c. 2×10^{13} cm^{-3} for $T_a = 300$ K; c. 10^{16} cm^{-3} for $T_a = 773$ K; c. 10^{13} cm^{-3} for $T_a = 873$ K; c. 10^{8} cm^{-3} for $T_a = 923$ K). Is this behavior consistent with compensation by structural atomic vacancies? Justify.

(d) Assuming that compensation is negligible for the unannealed sample ($T_a = 300$ K), estimate the concentration of atomic vacancies for the samples annealed at 773 K, 873 K and 923 K.

(e) The temperature dependence of (Hall) mobility has two distinct regimes: low temperatures (below 100 K) and high temperatures (above 100 K). In the high temperature regime all the samples show similar behavior (decreasing sharply with temperature), but at low temperatures there is a strong dependence on the sample. Justify this behavior by assigning a main scattering mechanism to each regime.

4

Optical properties

4.1 Introduction

The presence of the bandgap, as well as of various levels of impurities and/or defects, are fundamental elements of the optical properties of semiconductors and the source of numerous applications. In this chapter we will present some essential aspects of the optical properties of semiconductors. We will focus especially on the absorption coefficient and on the recombination effects of charge carriers, both interband and mediated by levels of defects/impurities located in the bandgap.

A vast set of optical processes in materials is possible. Light arriving at the surface of a material can be reflected and/or transmitted. Very general relations can be derived from the laws of classical electromagnetism. This leads to the well-known Fresnel[1] equations for the reflection coefficient R and the transmission coefficient T. Transmitted light is subject to a change in propagation speed (the well-known refraction process) and the refractive index n of the material is defined as the ratio $n = c/v$ between the velocity c of light in vacuum and the velocity of light v in the material. The velocity of light in the material is in principle dependent on the frequency ω ($v = v(\omega)$), and the corresponding dispersion effect arises as a consequence. It is often convenient to write the refractive index as a function of the incident (angular) frequency ω and of the wavenumber k (which varies with the material and can be dependent on the frequency):

$$n = \frac{c}{v} = \frac{c}{\omega}k \tag{4.1}$$

where the known relation for the phase velocity $v = \omega/k$ was used. We also recall the basic result obtained in classical electrodynamics for the propagation speed of light in an insulating medium with electrical permittivity ϵ and magnetic permittivity μ:

$$v = \frac{1}{\sqrt{\epsilon\,\mu}}, \tag{4.2}$$

[1]Augustin-Jean Fresnel (1788-1827).

DOI: 10.1201/9781003587736-4

from where the corresponding expression for the refractive index is derived:

$$n = \frac{c}{v} = \frac{\epsilon\mu}{\epsilon_0\mu_0} = \sqrt{\mu_r\epsilon_r}, \tag{4.3}$$

where $\epsilon_r = \epsilon/\epsilon_0$ is the relative electrical permittivity and $\mu_r = \mu/\mu_0$ is the relative magnetic permittivity (ϵ_0 and μ_0 being the corresponding electric and magnetic permittivities in vacuum). For non-magnetic materials, the relative magnetic permittivity is negligible ($\mu_r \sim 1$) and the following important relation is obtained:

$$n = \sqrt{\epsilon_r} \tag{4.4}$$

In the course of transmission, light can be absorbed, causing excitations in the solid, namely electron-hole pairs created by light with the appropriate frequency (typically in the visible zone, as we shall see below). These electron-hole pairs can subsequently recombine, emitting the corresponding (photo)luminescence, which provides a wealth of information about the material. However, it should be noted that whereas the incident light is transmitted in a certain direction, in the process of spontaneous emission light is emitted in all directions. This results in a decrease in the intensity I of the transmitted beam. The relative decrease in intensity dI/I is, in first order, simply proportional to the traveled distance dx:

$$dI = -\alpha\, I\, dx \Rightarrow I = I_0 \exp(-\alpha\, x) \tag{4.5}$$

α represents the absorption coefficient, which depends on the details of the electronic structure of the material and is therefore strongly dependent on the frequency $\alpha = \alpha(\omega)$. The result 4.5 is known as *Beer-Lambert law.*[2]

The intensity of the electromagnetic wave is simply proportional to the square of the amplitude of the electromagnetic wave ($I \propto |E|^2$), so Beer-Lambert law implies an exponential reduction in the amplitude of the propagating electric field, leading to solutions of the type (for a wave propagating in the x direction, written in the usual complex form):

$$\mathbf{E}(x,t) = \mathbf{E_0}\, e^{-\kappa'\, x} e^{j(k\, x - \omega\, t + \phi_0)} \tag{4.6}$$

where κ' is related to the decrease in the amplitude of the electric field as the wave progresses through the medium. This solution implies a decrease of intensity with distance:

$$I \propto |E|^2 \propto e^{-2\,\kappa'\, x} \tag{4.7}$$

the absorption coefficient, defined by eq. 4.5, is therefore:

$$\alpha = 2\,\kappa' \tag{4.8}$$

[2]August Beer (1825-1863); Johann Lambert (1728-1777).

4.2 Phenomenological description of absorption

The electric field 4.6 can also be written in a way similar to a plane wave without attenuation, taking into account that $-\kappa' = j^2\kappa'$ (where $j = \sqrt{-1}$):

$$\mathbf{E}(x,t) = \mathbf{E_0}\, e^{j\left[(k+j\kappa')x-\omega\, t+\phi_0\right]} = \mathbf{E_0}\, e^{j\left(\tilde{k}x-\omega\, t+\phi_0\right)} \tag{4.9}$$

We can thus introduce, in the ordinary plane wave solution, the effects due to wave absorption in the medium if we define a complex wave vector, whose real part simply corresponds to the wave number $k = 2\pi/\lambda$ and whose imaginary part corresponds to the attenuation coefficient κ':

$$\tilde{k} = k_r + j\, k_i = \frac{2\pi}{\lambda} + j\,\kappa' \tag{4.10}$$

Given relation 4.1 between the refractive index and the wave number k, it is useful to define a complex refractive index $\tilde{n}$ in the context of propagation with absorption:

$$\tilde{n} = \frac{c}{\omega}\tilde{k} = n_r + j\, n_i = \frac{c}{\omega}k_r + j\,\frac{c}{\omega}\, k_i = \frac{c}{v} + j\,\frac{c}{\omega}\,\kappa' = n + j\kappa \tag{4.11}$$

where the *extinction coeficient* κ is:

$$\kappa = \frac{c}{\omega}\kappa' = \frac{c\alpha}{2\omega} = \frac{\lambda_0\alpha}{4\pi} \tag{4.12}$$

where eq. 4.8 was used and where $\lambda_0 = 2\pi c/\omega$ is the wavelength in vacuum of the corresponding electromagnetic wave with angular frequency ω.

The relation 4.4 between the refractive index and the electrical permittivity of a non-magnetic material also allows us to define a complex electrical permittivity $\tilde{\epsilon} = \epsilon_r + j\epsilon_i$:

$$\tilde{n}^2 = \tilde{\epsilon} \tag{4.13}$$

where

$$\begin{aligned}\epsilon_r &= n^2 - \kappa^2 \\ \epsilon_i &= 2\, n\,\kappa\end{aligned} \tag{4.14}$$

and also

$$\begin{aligned}n &= \frac{1}{\sqrt{2}}\left[\epsilon_r + \left(\epsilon_r^2 + \epsilon_i^2\right)^{\frac{1}{2}}\right]^{\frac{1}{2}} \\ \kappa &= \frac{1}{\sqrt{2}}\left[-\epsilon_r + \left(\epsilon_r^2 + \epsilon_i^2\right)^{\frac{1}{2}}\right]^{\frac{1}{2}}\end{aligned} \tag{4.15}$$

For a small absorption coefficient ($\alpha = 2\kappa << n$), we simply get (from eq. 4.14):

$$
\begin{aligned}
n &= \sqrt{\epsilon_r} \\
\kappa &= \frac{\epsilon_i}{2n}
\end{aligned}
\tag{4.16}
$$

It should be noted that, although the refractive index expresses the dispersive effects and the absorption coefficient expresses the dissipative effects, eq. 4.15 establish a connection between these two types of effects through a single complex function $\tilde{\epsilon}(\omega)$. It is possible to establish very general relations (*Kramers-Kronig relations*[3]) between the real and imaginary parts of this complex function, in order to completely determine one as a function of the other: dispersion and dissipation are thus closely interlinked.[4]

Finally, it is possible to relate the complex refractive index to reflectivity[5]:

$$R = \left|\frac{\tilde{n}-1}{\tilde{n}+1}\right|^2 = \frac{(n-1)^2+\kappa^2}{(n+1)^2+\kappa^2} \tag{4.17}$$

4.3 Interband absorption

4.3.1 Direct bandgap and indirect bandgap

The presence of the bandgap has a very strong impact on the absorption coefficient of the semiconductor: as soon as the energy of the incident light exceeds the value of the bandgap, the material is suddenly able to absorb light in order to excite an electron from the valence band to the conduction band. A very important aspect is that the linear momentum of the photon is in general negligible when compared to the linear momentum of the electron (which is of the order of – but not identical to – $\hbar k$, k being the crystal momentum of the Bloch wavefunction). The absorption of a photon by an electron therefore only allows vertical transitions in the band structure. This establishes an important qualitative difference between direct bandgap semiconductors, where the minimum energy of the conduction band and the maximum energy of the valence band occur for the same k, and indirect bandgap semiconductors, where the minimum and maximum occur for different values of k.

The transition between the minimum energy of the conduction band and the maximum energy of the valence band is permitted for direct bandgap semiconductors, but in the case of indirect bandgap semiconductors it can only occur if the change in the electron's crystal momentum is compensated for by the lattice, typically by the emission/absorption of a phonon. This process

[3] Ralph Kronig (1904-1995); Hendrik Kramers (1894-1952).

[4] For further developments, see J. D. Jackson, *Classical electrodynamics*, 3rd edition, pp. 330-335.

[5] Cf. Mark Fox, *Optical Properties of Solids*, 2nd edition, Appendix A.

involves the concurrence of three entities (electron, photon and phonon), and is therefore much less likely. The absorption coefficient of indirect bandgap semiconductors is therefore typically much lower than that of direct bandgap semiconductors for energies close to the bandgap.

4.3.2 Transition rate for direct bandgap semiconductors

The absorption coefficient for the interband process in a direct bandgap semiconductor is simply proportional to the transition rate driven by the electromagnetic excitation. The transition rate $W_{i\to f}$ between an initial quantum state $|i\rangle$ and a final state $|f\rangle$, for direct absorption in semiconductors, can be calculated using *Fermi golden rule* of perturbation theory:

$$W_{i\to f} = \frac{2\pi}{\hbar} |M|^2 g(\hbar\omega) \tag{4.18}$$

where $|M| = \langle f| H' |i\rangle$ is the matrix element of the perturbation and $g(\hbar\omega)$ is the corresponding joint density of states. The matrix element can be calculated in the *dipolar approximation* $H' = -\mathbf{p}\cdot\varepsilon$, where $\mathbf{p} = -e\mathbf{r}$ is the dipole moment and ε is the electric field, which we assume to be polarized in the x direction:

$$\varepsilon = \varepsilon_0 e^{i(\pm\mathbf{k}\cdot\mathbf{r}-\omega t)}\hat{e}_x \tag{4.19}$$

The initial state $|i\rangle$ and the final state $|f\rangle$ correspond to Bloch functions:

$$\begin{aligned} |i\rangle &= \frac{1}{\sqrt{V}} u_i(\mathbf{r}) e^{i(i\mathbf{k_i}\cdot\mathbf{r})} \\ |f\rangle &= \frac{1}{\sqrt{V}} u_f(\mathbf{r}) e^{i(i\mathbf{k_f}\cdot\mathbf{r})} \end{aligned} \tag{4.20}$$

The matrix element is therefore:

$$M \propto \int u_f^*(\mathbf{r})\, x\, u_i(\mathbf{r}) d^3\mathbf{r} \tag{4.21}$$

This matrix element is dependent on the details of the band structure, but is not dependent on the energy of the incident photon.

On the other hand, the transition rate depends on the joint density of states associated with the transition. Assuming parabolic bands $E_C(k) = E_g + \hbar^2 k^2/2m_e^*$ and $E_V(k) = -\hbar^2 k^2/2m_h^*$, energy conservation in the transition implies that:

$$\hbar\omega = E_g + \frac{\hbar^2 k_e^2}{2\, m_e^*} + \frac{\hbar^2 k_h^2}{2\, m_h^*} = E_g + \frac{\hbar^2 k^2}{2\, m_e^*} + \frac{\hbar^2 k^2}{2\, m_h^*} \tag{4.22}$$

where we considered that the transition is vertical in the band structure, implying $k_e = k_h$. This relation can also be written as a single parabolic band:

$$\hbar\omega - E_g = \frac{\hbar^2 k^2}{2}\left(\frac{1}{m_h^*} + \frac{1}{m_e^*}\right) = \frac{\hbar^2 k^2}{2\,\mu} \tag{4.23}$$

where the effective mass μ is defined through the relation

$$\frac{1}{\mu} = \frac{1}{m_h^*} + \frac{1}{m_e^*} \tag{4.24}$$

The *joint density of states* $g(\hbar\omega)$ associated with the transition thus simply corresponds, for $\hbar\omega > E_g$, to the density of states of a parabolic band $\hbar^2 k^2/(2\mu)$, and is zero for $\hbar\omega < E_g$:

$$g(\hbar\omega) = \begin{cases} 0 & , \quad \hbar\omega < E_g \\ \frac{1}{2\pi^2}\left(\frac{2\mu}{\hbar^2}\right)^{3/2}(\hbar\omega - E_g)^{1/2} & , \quad \hbar\omega > E_g \end{cases}$$

The absorption coefficient thus shows, for direct bandgap semiconductors, a characteristic dependence on $\sqrt{\hbar\omega - E_g}$:

$$\alpha(\hbar\omega) \propto W_{i\to f} \propto |M|^2 g(\hbar\omega) \propto (\hbar\omega - E_g)^{\frac{1}{2}}, \hbar\omega > E_g \tag{4.25}$$

Figure 4.1 shows experimental data for GaAs (a direct bandgap semiconductor), taken from the literature, where the abrupt variation in the absorption coefficient can be seen for energies close to the bandgap, as well as the variation of the bandgap with temperature.

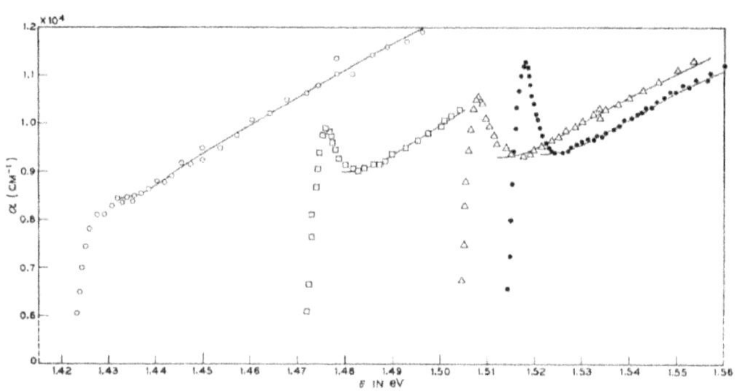

FIG. 3 Exciton absorption in GaAs; ○ 294°K, □ 186°K, △ 90°K, ● 21°K.

FIGURE 4.1
Absorption coefficient of GaAs, as a function of energy, for energies close to the bandgap value and for various temperatures (○ 294 K; □ 186 K; △ 90 K; ● 21 K). The peak observed near the bandgap is an excitonic effect that will be discussed later (Figure reproduced with permission from M. D. Sturge, *Optical Absorption of Gallium Arsenide between 0.6 and 2.75 eV*, Phys. Rev. **127**, 768 (1962). *Copyright* (1962) of the American Physical Society).

The model we have just presented provides a basic description of light absorption in direct bandgap semiconductors. More complete models also include the following aspects not covered in this model:

1. the Coulomb attraction between conduction electrons and holes can lead to an increase in the absorption rate due to the formation of excitons; this effect is more intense for high bandgap semiconductors at low temperatures;
2. the presence of impurities/defects with energy levels in the bandgap leads to light absorption for photons with energy $\hbar\omega < E_g$;
3. the parabolic approximation fails for $k >> 0$, implying in these cases a detailed knowledge of the $E(k)$ band structure in order to calculate the absorption coefficient.

In the case of indirect bandgap semiconductors, as mentioned above, the absorption of light with energy $\hbar\omega$ larger than the energy of the bandgap is only possible if there is simultaneous excitation of a phonon that provides the electron with the crystal momentum necessary for the transition. In the presence of a phonon with energy $\hbar\Omega$ and (crystal-) momentum $\hbar\mathbf{q}$, the conservation of energy in the process is as follows

$$E_f = E_i + \hbar\omega \pm \hbar\Omega \tag{4.26}$$

and the conservation of (crystal) momentum is:

$$\hbar\mathbf{k_f} = \hbar\mathbf{k_i} \pm \hbar\mathbf{q} \tag{4.27}$$

It can be shown that, in the case of an indirect bandgap semiconductor, the absorption rate in the vicinity of the absorption threshold has a quadratic dependence on the incident photon energy:[6]

$$\alpha(\hbar\omega) \propto (\hbar\omega - E_g \mp \hbar\Omega)^2 \, , \hbar\omega > E_g \tag{4.28}$$

The energy dependence of the absorption coefficient at the absorption threshold thus provides a simple means of distinguishing the nature of the bandgap: direct or indirect. Furthermore, the absorption coefficient for indirect transitions is much lower than that for direct transitions, at the absorption threshold. However, when the incident light reaches sufficient energy, the direct transitions (corresponding to $\hbar\omega >> E_g$) can be excited, which corresponds to a drastic increase in the absorption coefficient.

4.4 Excitons

The electron-hole pair formed through optical excitation can form a pair bonded by electrostatic attraction – an *exciton.* This exciton is a hydrogenic system similar to positronium (positron-electron system bound by

[6]Cf. Peter Y. Yu e Manuel Cardona, *Fundamentals of Semiconductors: Physics and Materials Properties*, p. 263.

electrostatic attraction). The attraction between the electron and the hole is strongly conditioned by the electrostatic shielding inside the material, described by the dielectric constant. This leads to two limiting cases:

1. *bound excitons* or *Frenkel excitons*[7], typical of materials with a low dielectric constant (molecular materials), where the electron-hole pair is strongly bound (with ionization energies of the order of 0.1–1 eV) and can be observed at room temperature; these excitons are generally localized in the material;
2. *free excitons* or *Wannier-Mott excitons*[8], typical of materials with high dielectric constant (such as semiconductors), where the binding energy of the electron-hole pair is consequently very low (of the order of 10 meV) and the radius of the exciton is large; these excitons are therefore only observed at low temperatures and can move freely through the crystal.

4.4.1 Free excitons

Free excitons can be adequately described by the same hydrogenic model used to describe the *shallow* impurities (section 2.2.1). The binding energy can be easily estimated using Bohr model, modified by the following considerations:

1. the effect of the crystal potential implies that the electron and hole forming the exciton are described by the effective mass of conduction electrons and holes in the crystal lattice, respectively;
2. the exciton thus behaves like a hydrogenic system; however, unlike what happens in ordinary hydrogenic systems, where the nucleus has a much higher mass than the electron, in the case of the exciton the electron and the hole have similar masses; the description of the dynamics of the exciton in the center-of-mass reference system thus requires that the reduced mass of the system be considered:[9]

$$\frac{1}{\mu} = \frac{1}{m_h^*} + \frac{1}{m_e^*} \tag{4.29}$$

   where m_e^* and m_h^* are the effective masses of the electron and of the hole, respectively.
3. in addition, the Coulomb attractive potential between the electron and the hole is strongly diminished compared to the value in vacuum, due to the presence of the shielding of the charges present, translated into the dielectric constant ϵ of the material.

---

[7] Yakov Frenkel (1894-1952).

[8] Gregory Wannier (1911-1983); Nevill Mott (1905-1996).

[9] Cf. B. H. Bransden and C. J. Joachain, *Physics of Atoms and Molecules*, 2nd edition, p. 109.

The energy levels of the exciton are thus again given by eq. 2.29:

$$E_n = \frac{1}{n^2}\frac{\epsilon_0^2}{\epsilon^2}\frac{\mu}{m_0}E_0 = \frac{R_X}{n^2} \tag{4.30}$$

where $E_0 = -e^4 m_0/\left(2(4\pi\epsilon_0)^2\hbar^2\right) = -13.6$ eV, m_0 is the mass of the free electron and ϵ_0 is the electrical permittivity of vacuum.
Similarly, the exciton radius can be estimated from eq. 2.27:

$$r_n = n^2\frac{\epsilon}{\epsilon_0}\frac{m_0}{m^*}r_0 = n^2 a_X \tag{4.31}$$

where $r_0 = 4\pi\epsilon_0\hbar^2/(e^2 m_0) = 0.53$ Å is the Bohr radius of the ground state of hydrogen. This model adequately describes the binding energies of free excitons.

4.4.1.1 Absorption by free excitons

A necessary condition for the formation of an exciton is that the electron velocity and the hole velocity are the same:

$$v_e = v_h \Rightarrow \frac{1}{\hbar}\left.\frac{\partial E}{\partial k}\right|_e = \frac{1}{\hbar}\left.\frac{\partial E}{\partial k}\right|_h \tag{4.32}$$

This condition implies that the conduction band gradient and the valence band gradient are identical at the point in the Brillouin zone where the transition occurs. This happens naturally in the transition $\hbar\omega = E_g$ in direct bandgap semiconductors, where both gradients are zero at $k = 0$. Excitons are thus created in direct transitions at $k = 0$. The energy needed to create the exciton corresponds to the value of the bandgap subtracted from the exciton's binding energy:

$$E_n = E_g - \frac{R_X}{n^2} \tag{4.33}$$

The peak observed for energies close to the bandgap in figure 4.1 illustrates the effect of excitons on the absorption coefficient. An important aspect is that the formation of excitons generally requires very pure samples (and is sometimes taken as a signature of purity). Electrons and holes released by impurities shield the Coulomb potential and tend to prevent exciton formation. Excitons are usually not observed in highly doped semiconductors or in metals.

4.5 Absorption by free charge carriers

We now consider the absorption of light due to free carriers in a semiconductor. This corresponds to analyzing the behavior of electromagnetic waves in materials with a non-negligible conductivity. Even a small conductivity will lead to the dissipation of energy through the Joule effect.[10] This will have consequences for the propagation of electromagnetic waves, leading to an attenuation of their amplitude and possible extinction. We therefore expect the resulting wave equation to contain a dissipative term, which will depend on conductivity.

To obtain the wave equation, we start from Faraday law:[11]

$$\nabla \times \mathbf{E} = -\frac{\partial \mathbf{B}}{\partial t} \tag{4.34}$$

We calculate the rotational of this equation, assuming that the total charge density ρ is zero (as is usually the case in the presence of electric currents, since the material remains overall neutral):

$$\begin{aligned} \nabla \times (\nabla \times \mathbf{E}) &= -\frac{\partial}{\partial t}(\nabla \times \mathbf{B}) \Leftrightarrow \\ \Leftrightarrow -\nabla^2 \mathbf{E} + \nabla(\nabla \cdot \mathbf{E}) &= -\mu \frac{\partial \mathbf{J}}{\partial t} - \mu\epsilon \frac{\partial^2 \mathbf{E}}{\partial t^2} \Leftrightarrow \\ \Leftrightarrow \nabla^2 \mathbf{E} &= \sigma\mu \frac{\partial \mathbf{E}}{\partial t} + \epsilon\mu \frac{\partial^2 \mathbf{E}}{\partial t^2} \end{aligned} \tag{4.35}$$

In addition to Ohm law $\mathbf{J} = \sigma\mathbf{E}$ and Gauss law[12] in the case of $\rho = 0$, we also used Ampère-Maxwell law,[13] which is written in this case as:

$$\nabla \times \mathbf{B} = \mu\mathbf{J} + \mu\epsilon \frac{\partial \mathbf{E}}{\partial t} \tag{4.36}$$

The equation we have just obtained corresponds, as expected, to a damped wave equation, with a damping term $\sigma\mu\partial\mathbf{E}/\partial t$ proportional to the conductivity.

4.5.1 Dependence with conductivity and frequency

We now verify that eq. 4.9 is a solution of the damped wave equation (eq. 4.35):

$$\nabla^2 \mathbf{E} = -\tilde{k}^2\, e^{j(\tilde{k}x - \omega\, t + \phi_0)} \tag{4.37}$$

$$\sigma\mu \frac{\partial \mathbf{E}}{\partial t} = -j\omega\sigma\mu\, e^{j(\tilde{k}x - \omega\, t + \phi_0)} \tag{4.38}$$

$$\epsilon\mu \frac{\partial^2 \mathbf{E}}{\partial t^2} = -\omega^2\epsilon\mu\, e^{j(\tilde{k}x - \omega\, t + \phi_0)} \tag{4.39}$$

[10] James Joule (1818-1889).

[11] Michael Faraday (1791-1867).

[12] Johann Carl Friedrich Gauss (1777-1855).

[13] André-Marie Ampère (1775-1836); James Maxwell (1831-1879).

Expression 4.9 is a solution if

$$\tilde{k}^2 = j\omega\sigma\mu + \omega^2\epsilon\mu \tag{4.40}$$

Or, taking into account that $\tilde{n} = \frac{c}{\omega}\tilde{k}$ and that $\epsilon = \epsilon_0\epsilon_r$, $\mu = \mu_0\mu_r$:

$$\tilde{n}^2 = j\frac{\sigma\mu_r}{\omega\epsilon_0} + \epsilon_r\mu_r \tag{4.41}$$

This equation, together with $\tilde{n} = n_r + j\,n_i$, allows us to obtain the real component and the imaginary component of the complex refractive index. In fact, we have:

$$\tilde{n}^2 = (n_r^2 - n_i^2) + j\,2\,n_r\,n_i \tag{4.42}$$

This implies that:

$$\begin{aligned} n_r^2 - n_i^2 &= \epsilon_r\mu_r \\ 2\,n_r\,n_i &= \frac{\sigma\mu_r}{\omega\epsilon_0} \end{aligned} \tag{4.43}$$

This leads us to the biquadratical equation in n_r:

$$n_r^4 - \epsilon_r\mu_r\,n_r^2 - \frac{\sigma^2\mu_r^2}{4\omega^2\epsilon_0^2} = 0 \tag{4.44}$$

and the only positive solution of this equation is:

$$n_r = \left(\frac{\epsilon_r\mu_r}{2}\right)^{\frac{1}{2}}\left[1 + \left(1 + \left(\frac{\sigma}{\omega\epsilon}\right)^2\right)^{\frac{1}{2}}\right]^{\frac{1}{2}} \tag{4.45}$$

From here we immediately obtain, from the relation $(n_r - n_i)^2 = \epsilon_r\mu_r$, the corresponding expression for n_i:

$$n_i = \left(\frac{\epsilon_r\mu_r}{2}\right)^{\frac{1}{2}}\left[-1 + \left(1 + \left(\frac{\sigma}{\omega\epsilon}\right)^2\right)^{\frac{1}{2}}\right]^{\frac{1}{2}} \tag{4.46}$$

The typical penetration length of the electromagnetic wave is therefore (Cf. eq. 4.11):

$$\frac{1}{\kappa'} = \frac{1}{k_i} = \frac{c}{n_i\omega} \tag{4.47}$$

and the absorption coefficient $\alpha = 2\kappa'$ is:

$$\alpha = 2\kappa' = \frac{2n_i\omega}{c} \tag{4.48}$$

4.5.1.1 Relation with AC conductivity

We recall that the dynamics of conduction electrons is adequately described, in the presence of dispersion mechanisms associated with the average time τ, by eq. 3.23:

$$m^* \frac{d\mathbf{v}}{dt} = -e\mathbf{E} - m^* \frac{\mathbf{v}}{\tau} \tag{4.49}$$

This equation is related to a current density proportional to the applied electric field, according to Ohm law:

$$\mathbf{J} = \sigma_0 \mathbf{E} = nq\mathbf{v} \tag{4.50}$$

where $\sigma_0 = n\, e^2 \tau / m^*$ is the static (or DC) conductivity of the material.

In the presence of an electromagnetic wave described by the eq. 4.9, it is still possible to use eq. 4.50 (Ohm law), while simultaneously describing the effects of attenuation due to conductivity and propagation of the wave in the conductive medium. This results in the following expression for AC conductivity:

$$\sigma(\omega) = \frac{\sigma_0}{1 + \omega^2 \tau^2} + i\sigma_0 \frac{\omega\tau}{1 + \omega^2 \tau^2} = \sigma_0 \exp\left[i \arctan(\omega\tau)\right] \tag{4.51}$$

The conductivity in eq. 4.51 is expressed by a complex number which describes not only the relation between the moduli of the applied electric field and the corresponding current density, but also the respective phase difference $\arctan(\omega\tau)$. Note that at low frequencies the real term dominates (leading to DC conductivity, σ_0) and at high frequencies the imaginary term dominates.

4.5.1.2 Limit cases: insulators and metals

The description we have obtained for waves propagating in a medium with conductivity $\sigma = 0$ allows us to write, from eqs. 4.45 and 4.46:

$$n_r = \sqrt{\mu_r \epsilon_r} \tag{4.52}$$

$$n_i = 0 \tag{4.53}$$

In this limiting case, as expected, there is no absorption, light propagates in the medium with a refractive index $n = \sqrt{\mu_r \epsilon_r}$, and the electric and magnetic fields propagate with the same phase.

We now consider the opposite limit, where the conductivity is very high. This implies considering the following limit in eqs. 4.45 and 4.46:

$$\frac{\sigma}{\omega\epsilon} >> 1 \tag{4.54}$$

The limit of high conductivity is therefore frequency dependent. A good conductor in one spectral region may not be so in other frequency regions. In this approximation, we obtain:

$$n_r = \sqrt{\frac{\mu_r \sigma}{2\epsilon_0 \omega}} \tag{4.55}$$

$$n_i = n_r = \sqrt{\frac{\mu_r \sigma}{2\epsilon_0 \omega}} \tag{4.56}$$

The damping of the electromagnetic wave is described, as we have seen, by the imaginary component of the refractive index, and the typical length for wave attenuation is given by:

$$\delta = \frac{1}{\kappa'} = \frac{1}{k_i} = \frac{c}{\omega\, n_i} = \frac{c}{\omega}\sqrt{\frac{2\epsilon_0 \omega}{\mu_r \sigma}} = \frac{1}{\sqrt{\pi\, \mu\, \sigma\, f}} \tag{4.57}$$

while the absorption coefficient is given by:

$$\alpha = 2\kappa' = 2\sqrt{\pi\, \mu\, \sigma\, f} \tag{4.58}$$

Thus, in good conductors, the higher the frequency f of the propagating wave, the larger the attenuation. This leads to rapid damping of the electromagnetic waves inside the conductor – electromagnetic waves in conductors therefore only propagate in a surface layer with a typical thickness δ. This effect is usually called *skin effect* and leads to a progressive increase in the resistance of the conductors with frequency.

4.5.1.3 High frequency limit

The high frequency limit is also worth analyzing in detail. Note that in this case the conductivity is dominated by the imaginary term in eq. 4.51, so the best way to analyze the impact of high frequencies on the absorption coefficient $\alpha = 2\omega\, n_i/c$ (eq. 4.48), is to directly calculate n_i from eq. 4.41. In the common case where the material is non-magnetic ($\mu_r \sim 1$), this expression reduces to:

$$\tilde{n}^2 = \epsilon_r + j\frac{\sigma}{\omega\epsilon_0} \tag{4.59}$$

and, taking into account the complex expression of the AC conductivity (eq. 4.51), we get:

$$\tilde{n}^2 = \left(\epsilon_r - \frac{\sigma_0}{\omega\epsilon_0}\frac{\omega\tau}{1+\omega^2\tau^2}\right) + i\left(\frac{1}{\omega\epsilon_0}\frac{\sigma_0}{1+\omega^2\tau^2}\right) \tag{4.60}$$

Using eq. 4.42 once again and taking into account that $\sigma_0 = n\, e^2\, \tau/m^*$, we get:

$$\begin{aligned} n_r^2 &= \epsilon_r - \frac{n\, e^2 \tau^2}{m^* \epsilon_0\, (1+\omega^2\tau^2)} + n_i^2 \\ 2\, n_r\, n_i &= \frac{n\, e^2 \tau}{m^* \omega \epsilon_0\, (1+\omega^2\tau^2)}, \end{aligned} \tag{4.61}$$

This leads us, in the high-frequency limit ($\omega\tau >> 1$), to:

$$\begin{aligned} n_r &\sim \sqrt{\epsilon_r} \\ n_i &\sim \frac{n\,e^2}{2m^*\omega^3\tau\epsilon_0\sqrt{\epsilon_r}}, \end{aligned} \tag{4.62}$$

The absorption coefficient $\alpha = 2\omega\, n_i/c$ (eq. 4.48) therefore becomes:

$$\alpha = \frac{2\omega\, n_i}{c} \sim \frac{n\,e^2}{m^*\omega^2\tau\, c\epsilon_0\sqrt{\epsilon_r}}, \tag{4.63}$$

The absorption coefficient therefore decreases dramatically with frequency, which implies that light is progressively less absorbed as the frequency increases. We can appreciate more precisely what is happening by analyzing the frequency dependence of the real component of the refractive index, which reflects the propagation of the electromagnetic wave. Returning to eq. 4.61, n_r is expressed as:

$$n_r^2 = \epsilon_r - \frac{n\,e^2\tau^2}{m^*\epsilon_0\,(1+\omega^2\tau^2)} + n_i^2 \tag{4.64}$$

However, n_r depends on frequency either by the term in ω^{-2} or by the term in n_i^2 (which has a dependence on frequency in ω^{-6}, given the dependence of n_i on ω^{-3} expressed in eq. 4.62). The dependence of n_r on frequency for high frequencies will be dominated by the term in ω^{-2}, as follows:

$$n_r^2 \sim \epsilon_r - \frac{n\,e^2}{m^*\epsilon_0\omega^2} = \epsilon_r\left[1 - \left(\frac{\omega_p}{\omega}\right)^2\right] \tag{4.65}$$

where we have defined the frequency ω_p as

$$\omega_p = \sqrt{\frac{ne^2}{m^*\epsilon_0\epsilon_r}} \tag{4.66}$$

Equation 4.65 shows the behavior of the refractive index for high frequencies. In the limit where $\omega >> \omega_p$, the refractive index progressively tends toward $\sqrt{\epsilon_r}$; as the frequency ω decreases, the term in ω_p/ω becomes progressively more important, leading to a decrease in the refractive index (and consequent increase in the speed of propagation).

The ω_p frequency that conditions the evolution of the refractive index with frequency is called the *plasma frequency* and reflects the natural oscillation frequency of the free electron gas after being excited by an electric field. Expression 4.65 clearly loses meaning for $\omega \sim \omega_p$ (corresponding to a zero refractive index). For frequencies ω lower than the plasma frequency ω_p, expression 4.65 corresponds to a purely imaginary refractive index, implying that no propagation occurs (complete absorption/reflection). Metals and highly doped semiconductors therefore tend to absorb/reflect all radiation for frequencies below the plasma frequency, and are transparent for frequencies above ω_p. In metals, ω_p typically corresponds to ultraviolet radiation, so metals absorb/reflect radiation in the visible zone.

4.6 Recombination

The interband absorption of light by a semiconductor leads to the creation of a concentration n of conduction electrons higher than the thermal equilibrium concentration n_{eq}:

$$n = n_{eq} + \Delta n \tag{4.67}$$

where Δn is the excess concentration. In the interband process, this concentration of excess conduction electrons naturally corresponds to a concentration of excess holes in the valence band: $\Delta p = \Delta n$. This deviation from thermodynamic equilibrium caused by the creation of excess charge carriers at a rate G is compensated for by the recombination of the charge carriers, with a characteristic recombination time τ. In a first approximation, the recombination is simply proportional to the excess concentration Δn, and the kinetics of the process is therefore characterized by the equation:

$$\frac{d\Delta n}{dt} = G - \frac{\Delta n}{\tau} \tag{4.68}$$

The joint occurrence of generation and recombination leads to an equilibrium value of the excess concentration:

$$\Delta n = G\tau \tag{4.69}$$

If the mechanism responsible for generation is turned off after the excess carrier concentration $\Delta n = G\tau$ is established, the solution of eq. 4.68 is a simple exponential (in the approximation where τ is constant):

$$\Delta n(t) = G\tau \exp\left(-\frac{t}{\tau}\right) \tag{4.70}$$

4.6.1 Interband recombination

Interband recombination is the inverse process of interband absorption, where a photon is emitted with the energy released by the conduction electron as it transits to the valence band. As in the absorption process, the photon is unable to carry any significant linear momentum of the electron, and this implies vertical transitions in the band structure. Transitions implying a change in the linear momentum of the electron, such as those in indirect bandgap semiconductors, require the intervention of lattice vibrations (phonons) and are therefore much less likely.

The interband recombination rate of a conduction electron with a hole in the valence band (or *bimolecular recombination rate*, r_B) depends on the concentration n of electrons in the conduction band and on the concentration p of holes in the valence band, as well as on their deviation from the respective values at thermal equilibrium, n_0 and p_0, respectively (where $n_0\,p_0 = n_i^2$,

TABLE 4.1
Bimolecular recombination coefficient for some semiconductors.

Material	B (cm^{-3} s^{-1})
GaAs	1.0×10^{-10}
InAs	2.1×10^{-11}
Si	1.1×10^{-14}
Ge	3.0×10^{-15}

according to the law of mass action). The bimolecular recombination rate can therefore be adequately described by:

$$r_B = B\,(n\,p - n_0\,p_0) = B\left(n\,p - n_i^2\right) \tag{4.71}$$

Typical values of the B coefficient can be found in table 4.1,[14] where the difference of 4 to 5 orders of magnitude between direct bandgap semiconductors (GaAs, InAs) and indirect bandgap semiconductors (Si, Ge) is evident.

In order to evaluate the time evolution of the concentration of excess charge carriers due to bimolecular interband recombination only (i.e. with generation rate $G = 0$), we consider:

$$\begin{aligned} n &= n_0 + \Delta\,n \\ p &= p_0 + \Delta\,p \end{aligned} \tag{4.72}$$

where

$$\Delta\,n = \Delta\,p \tag{4.73}$$

$$\frac{dn}{dt} = \frac{d(\Delta\,n)}{dt} = \frac{dp}{dt} = \frac{d(\Delta\,p)}{dt} \tag{4.74}$$

As a result, we get,

$$\frac{d(\Delta\,n)}{dt} = \frac{d\Delta\,p}{dt} = -r_B = -B\,(n\,p - n_0\,p_0) = -B\left(n\,p - n_i^2\right) \tag{4.75}$$

and, considering eqs. 4.72 and 4.73:

$$\frac{d(\Delta\,n)}{dt} = -B(n_0 + p_0)\Delta\,n - B\,(\Delta\,n)^2 \tag{4.76}$$

[14] Cf. M. Grundmann, *Physics of Semiconductors*, 2nd edition, p. 313.

Equation 4.76 can be easily integrated:

$$\frac{d(\Delta n)}{(n_0+p_0)\Delta n+(\Delta n)^2}=-Bdt \tag{4.77}$$

The integration can be performed by writing the denominator as:

$$\frac{1}{\Delta n\,(\Delta n+n_0+p_0)}=\frac{C_1}{\Delta n}+\frac{C_2}{\Delta n+n_0+p_0} \tag{4.78}$$

where

$$\begin{aligned} C_1 &= -C_2 \\ C_1 &= \frac{1}{n_0+p_0} \end{aligned} \tag{4.79}$$

The integration of eq. 4.77 leads to:

$$\frac{\ln\Delta n-\ln(\Delta n+n_0+p_0)}{n_0+p_0}=-B\,t \Leftrightarrow \frac{\Delta n}{(\Delta n+n_0+p_0)}=C\exp\left[-B\,(n_0+p_0)\,t\right] \tag{4.80}$$

The integration constant C can be found by establishing the initial condition $\Delta n(t=0)=\Delta n(0)$, leading to:

$$C=\frac{\Delta n(0)}{\Delta n(0)+n_0+p_0} \tag{4.81}$$

The general expression for $\Delta n(t)$ becomes:

$$\Delta n(t)=\frac{\Delta n(0)(n_0+p_0)}{(\Delta n(0)+n_0+p_0)\exp\left[B(n_0+p_0)t\right]-\Delta n(0)} \tag{4.82}$$

Taking into account that, in general, $\Delta n(0) << n_0+p_0$, this expression becomes the simple exponential 4.70 in the limit $\exp\left[B(n_0+p_0)t\right] >> \Delta n(0)/\left(n_0+p_0\right)$:

$$\Delta n(t)\sim\Delta n(0)\exp\left[-B(n_0+p_0)t\right] \tag{4.83}$$

In this case the characteristic time for recombination corresponds to :

$$\tau=\frac{1}{B(n_0+p_0)} \tag{4.84}$$

4.6.2 Auger recombination

Similarly to simple atomic systems, conduction electrons can also de-excite into the valence band by transferring the excess energy to another electron instead of emitting a photon (*Auger*[15] process). The Auger recombination process is therefore a non-radiative process that involves three bodies: the conduction electron, the hole and the third particle (conduction electron or

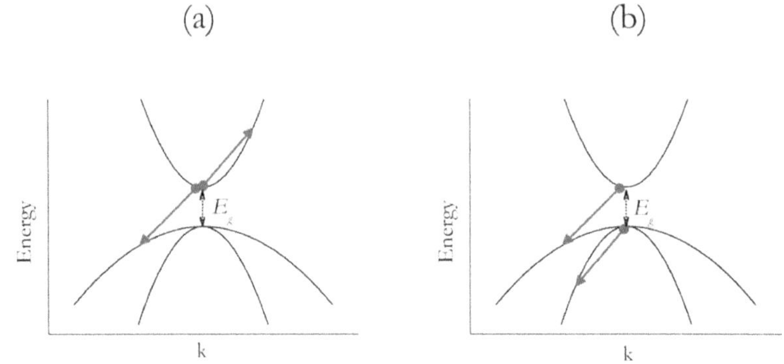

FIGURE 4.2
Schematic representation of Auger recombination. The energy resulting from the recombination of an electron with a hole can be transferred (a) to another electron in the conduction band; (b) or to another hole in the valence band.

hole) to which the excess energy is transferred. The two different possibilities are schematized in figure 4.2.

In the case where energy is transferred to a conduction electron, the recombination rate will now be proportional to n^2p, while in the case where energy is transferred to a hole, the recombination rate is proportional to p^2n. The kinetics of recombination, in the exclusive presence of Auger processes, is thus governed by the equation:

$$\frac{d\Delta n}{dt} = -C_n\left(n^2p - n_0^2p_0\right) - C_p\left(p^2n - p_0^2n_0\right) \tag{4.85}$$

where C_n and C_p are the Auger coefficients. The corresponding Auger recombination rate r_{Auger} is frequently approximated by the simpler expression:

$$r_{Auger} = \left(C_n n + C_p p\right)\left(np - n_0 p_0\right) \tag{4.86}$$

4.6.3 Recombination in the presence of defects/impurities

Since the physics of semiconductors is dominated by defects/impurities, there is naturally a significant impact of these on recombination processes. The recombination of an electron with a hole can thus occur at the impurity level. The rate of recombination depends essentially on the concentration of the impurity levels and on the position of the level in the bandgap. In the *Shockley-Read-Hall model*[16] (see problem 12), the corresponding

[15] Pierre Auger (1899-1993).
[16] William Shockley (1910-1989); William Read, Jr. (b. 1921); Robert Hall (1919-2016).

characteristic time τ can be written in the form

$$\tau = \frac{\tau_{p0}(n_0 + n_t) + \tau_{n0}(p_0 + p_t)}{n_0 + p_0} \tag{4.87}$$

where τ_{p0} and τ_{n0} are characteristic times associated with the capture of holes and electrons by impurities. The dependence of τ on the Fermi level is shown in figure 4.3.

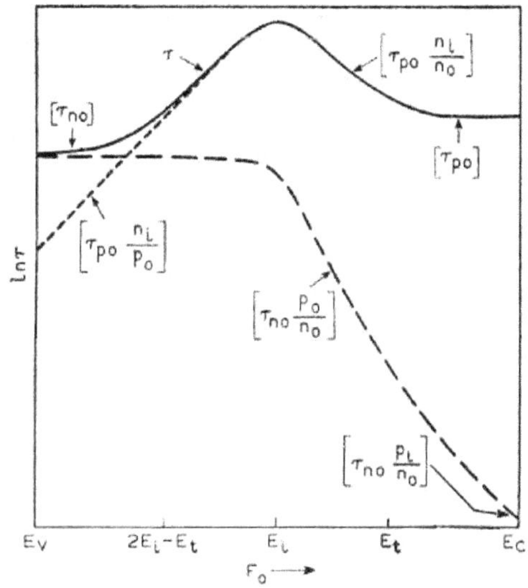

FIGURE 4.3
Dependence of the τ lifetime with the Fermi level, in the Shockley-Read-Hall model, (Figure reproduced with permission from W. Shockley and W. T. Read, Jr., *Statistics of the Recombinations of Holes and Electrons*, Phys. Rev. **87**, 835 (1952). *Copyright* (1952) of the American Physical Society).

4.6.4 Photoconductivity

The interband absorption of light leads, as we saw in previous sections, to an increase in charge carriers. This absorption process leaves a hole in the valence band for each electron promoted to the conduction band, which leads to an identical variation in their concentration (Δn and Δp, respectively):

$$\Delta n = \Delta p \tag{4.88}$$

As the charge concentrations of semiconductors are naturally low (especially in intrinsic or highly compensated semiconductors), these concentration variations due to the action of light (also sometimes called excess concentrations) can be very significant, with a corresponding impact on electrical conductivity. The variation in conductivity corresponding to the case of eq. 4.88 is simply (from eq. 3.19)

$$\Delta\sigma = e\,\Delta n\mu_e + e\,\Delta p\,\mu_h = e\,(\mu_e + \mu_h)\Delta n. \tag{4.89}$$

Naturally, the absorption of light and the generation of charge carriers in excess will be matched by an increase in the recombination rates U_n and U_p (for conduction electrons and holes, respectively). To a first approximation, these rates are approximately proportional to the excess concentrations:

$$\begin{aligned} U_n &= \frac{\Delta n}{\tau} \\ U_p &= \frac{\Delta p}{\tau} \end{aligned} \tag{4.90}$$

here τ is the recombination time (of the only interband process we are considering here).

In this approximation, where G is the optical generation rate (concentration of charge carriers generated per unit time, $G_n = G_p$ in the case of a single interband process), we can write approximately (Cf. eq. 3.59):

$$\frac{d\Delta n}{dt} = G - \frac{\Delta n}{\tau} \tag{4.91}$$

where the effects of diffusion have been disregarded for the time being. This is a reasonable approximation in the semiconductor volume subject to optical excitation as long as this excitation is constant. In the next section we will look at the effects of diffusion in more detail. As long as the optical excitation continues, there will be a simultaneous generation of charge carriers and their respective recombination, which, in the equilibrium situation of the two rates ($G = U \Rightarrow d\Delta n/dt = 0$), leads to an excess concentration given by:

$$(\Delta n)_{eq} = G\tau \tag{4.92}$$

4.6.5 Excess charge carrier profiles

When a beam of light with the appropriate wavelength hits the surface of a semiconductor, the process of light absorption begins, with generation of charge carriers (electrons in the conduction band and holes in the valence band). Concomitantly, the process(es) of recombination of these charge carriers also begin. At the same time, the concentration gradient of charge carriers implies a diffusion current from the surface to the interior of the material. Thus, even in the steady-state situation, where the concentration of excess charge carriers does not depend on time, it does show a dependence on depth (distance from the surface), dependent on the absorption coefficient, the characteristic recombination time and the diffusion constants. In order to systematically analyze the different contributions, we start by assuming that the absorption coefficient is very large and that all light is absorbed at the surface. We will subsequently consider the case where absorption/generation occurs in the bulk.

4.6.5.1 Generation at the surface

In the presence of diffusion and in the steady state ($d/dt = 0$), the continuity eqs. 3.59 reduce to eqs. 3.60:

$$\begin{aligned} \nabla^2 n &= -\frac{G_n - U_n}{D_n} \\ \nabla^2 p &= -\frac{G_p - U_p}{D_p} \end{aligned} \tag{4.93}$$

We consider the case of the depth profile of holes (the case of the depth profile of conduction electrons can be treated in a similar way) generated by a rate $G_p(x)$ that is only significant at the surface (corresponding to the case where light is all absorbed at a depth $\alpha^{-1} << L_p$, where α is the absorption coefficient and L_p is the diffusion length). In this case, an excess concentration Δp_0 of holes is generated at the surfac: $\Delta p_0 \equiv \Delta p(x = 0)$. The corresponding continuity equation for the excess concentration[17] Δp is written as (assuming that the surface is in the $x = 0$ plane, with diffusion into the interior of the semiconductor at $x > 0$):

$$D_p \frac{d^2 \Delta p}{dx^2} = -G_p(x) + \frac{\Delta p}{\tau_p} \tag{4.94}$$

where we have assumed $U_p = \Delta p/\tau_p$ for the recombination rate. For $x > 0$, $G(x) = 0$, and this equation becomes

$$\frac{d^2}{dx^2}\Delta p - \frac{1}{D_p \tau_p}\Delta p = 0 \tag{4.95}$$

Its general solution is

$$\Delta p(x) = C_1 \exp\left(x/L_p\right) + C_2 \exp\left(-x/L_p\right) \tag{4.96}$$

where the diffusion length L_p is:

$$L_p = \sqrt{D_p \tau_p} \tag{4.97}$$

The integration constants C_1 and C_2 are immediately determined from the boundary conditions $\Delta p(x = 0) = \Delta p_0$ and $\Delta p(x \to \infty) = 0$, resulting in

---

[17]Recall that $p = p_0 + \Delta_p$, where p_0 is the constant value corresponding to the concentration of holes in thermodynamic equilibrium in the absence of light. Thus, $\partial_t p = \partial_t (\Delta p)$ and $\partial_x p = \partial_x (\Delta p)$.

$C_1 = 0$ and $C_2 = \Delta p_0$. The profile of the excess concentration of holes thus takes the final form:

$$\Delta p(x) = \Delta p_0 \exp\left(-x/L_p\right) \tag{4.98}$$

The value of the excess concentration Δp_0 at the surface can be related to the total optical generation rate G_{tot}. The total optical generation rate G_{tot} is defined by the total amount Δp_{total} of holes generated, through eq. 4.92:

$$\Delta p_{total} = G_{tot}\tau_p \tag{4.99}$$

The number of holes generated varies with depth according to the profile 4.98, so the number of holes generated in the volume corresponding to the depths x and $x + dx$ is $\Delta p_0 \exp\left(-x/L_p\right) A dx$, where A is the surface area of the semiconductor. We have

$$G_{tot} = \frac{1}{\tau_p}\int_0^{\infty} \Delta p_0 \exp\left(-x/L_p\right) A dx \Leftrightarrow G_{tot} = \Delta p_0 A \frac{L_p}{\tau_p} \tag{4.100}$$

The total generation rate per unit area is therefore (considering the definition of the diffusion length in eq. 4.97):

$$\frac{G_{tot}}{A} = \Delta p_0 \sqrt{\frac{D_p}{\tau_p}} \tag{4.101}$$

4.6.5.2 Generation in the bulk

We now consider the case of the hole profile generated by a depth dependent generation rate $G_p(x)$, described by the exponential form

$$G_p(x) = G_0 \exp\left(-\alpha x\right), \tag{4.102}$$

This corresponds to a total generation rate:

$$G_{tot} = \int_0^{\infty} G(x)dx = \frac{G_0}{\alpha} \tag{4.103}$$

The corresponding continuity equation for the excess concentration Δp is now written (again assuming that the surface is in the $x = 0$ plane, with diffusion into the semiconductor occurring at $x > 0$):

$$\frac{d^2\Delta p}{dx^2} - \frac{\Delta p}{D_p\tau_p} = -\frac{G_0}{D_p}\exp\left(-\alpha x\right) \tag{4.104}$$

This equation is not homogeneous, but can be converted into homogeneous using the polynomial operator $d/dx + \alpha$, leading to a general solution of the form:

$$\Delta p(x) = C_1 \exp\left(x/L_p\right) + C_2 \exp\left(-x/L_p\right) + C_3 \exp\left(-\alpha x\right) \tag{4.105}$$

This general solution must obey eq. 4.104, leading to

$$C_3 = \frac{G_0 \tau_p}{1 - \alpha^2 L_p^2} \tag{4.106}$$

and to the following general solution of eq. 4.104:

$$\Delta p(x) = C_1 \exp(x/L_p) + C_2 \exp(-x/L_p) + \frac{G_0 \tau_p}{1 - \alpha^2 L_p^2} \exp(-\alpha x) \tag{4.107}$$

Again, the constants C_1 and C_2 depend on the boundary conditions, and it is common to apply the condition $\Delta p\,(x \to \infty) = 0$, which immediately leads to $C_1 = 0$. The constant C_2 can also be determined by assuming the existence of a surface recombination rate S (see problem 9):

$$-eS\Delta p(x=0) = -eD_p \left.\frac{\partial}{\partial x}\Delta p\right|_{x=0} \Leftrightarrow S\Delta p(x=0) = D_p \left.\frac{\partial}{\partial x}\Delta p\right|_{x=0} \tag{4.108}$$

leading to

$$C_2 = -\frac{S + \alpha D_p}{S + \frac{D_p}{L_p}} \frac{G_0 \tau_p}{1 - \alpha^2 L_p^2} \tag{4.109}$$

and to the general solution:

$$\Delta p(x) = \frac{G_0 \tau_p}{1 - \alpha^2 L_p^2} \left[\exp(-\alpha x) - \frac{S + \alpha D_p}{S + \frac{D_p}{L_p}} \exp(-x/L_p)\right] \tag{4.110}$$

Problems

1. Use eq. 4.14 to obtain eq. 4.15 for the refractive index and for the extinction index.
2. Obtain eq. 4.51 for the AC conductivity (see problem 6 in Chapter 3).
3. The transition rate for direct absorption in semiconductors can be calculated using the Fermi Golden Rule of perturbation theory:

$$W_{i \to f} = \frac{2\pi}{\hbar} |M|^2 g(\hbar\omega) \tag{4.111}$$

where $|M| = \langle f|\, H'\, |i\rangle$ is the matrix element of the transition and $g(\hbar\omega)$ is the corresponding joint density of states.

(a) Check that the matrix element, in the dipolar approximation $H' = -\mathbf{p} \cdot \epsilon$, obeys

$$M \propto \int u_f^*(\mathbf{r})\, x\, u_i(\mathbf{r}) d^3\mathbf{r} \tag{4.112}$$

where u_f and u_i are the periodic functions characteristic of Bloch functions. Assume that the electromagnetic wave is polarized in the x direction: $\varepsilon = \varepsilon_0 \exp i(\pm \mathbf{k} \cdot \mathbf{r} - \omega\, t)\hat{e}_x$.

(b) Obtain the joint density of states:

$$g(\hbar\omega) = \begin{cases} 0 & , \quad \hbar\omega < E_g \\ \frac{1}{2\pi^2}\left(\frac{2\mu}{\hbar^2}\right)^{3/2}(\hbar\omega - E_g)^{1/2} & , \quad \hbar\omega > E_g \end{cases}$$

where $\mu^{-1} = \mu_e^{-1} + \mu_h^{-1}$ is the electron/hole reduced mass.

4. *Absorption by free carriers.*

(a) Show that, for a material with conductivity σ (as well as electrical permittivity ϵ and magnetic permittivity μ), the wave equation for the electric field can be written as:

$$\nabla^2 \mathbf{E} = \sigma\mu \frac{\partial \mathbf{E}}{\partial t} + \epsilon\mu \frac{\partial^2 \mathbf{E}}{\partial t^2} \tag{4.113}$$

(b) Assuming a monochromatic damped wave solution $\mathbf{E}(x,t) = \mathbf{E_0}\, e^{-\kappa\, x} e^{j(k\, x - \omega\, t)}$, show that the complex refractive index $\tilde{n} = c/\omega\, \tilde{k} = c/v + ic\kappa/\omega$ becomes

$$\tilde{n}^2 = j\frac{\sigma\mu_r}{\omega\epsilon_0} + \epsilon_r\mu_r \tag{4.114}$$

(c) Show that for high frequencies ($\omega\tau >> 1$) the absorption coefficient $\alpha = 2\omega n_i/c$ of a non-magnetic material ($\mu_r = 1$) is written:

$$\alpha = \frac{ne^2}{c\epsilon_0 m^* \omega^2 \tau \sqrt{\epsilon_r}} \tag{4.115}$$

(d) Show that for high frequencies ($\omega\tau >> 1$) the real part of the refractive index is:

$$n_r^2 = \epsilon_r \left[1 - \left(\frac{\omega_p}{\omega}\right)\right]^2 \tag{4.116}$$

where $\omega_p = \sqrt{\frac{ne^2}{\epsilon m*}}$ is the *plasma frequency*. Discuss the cases $\omega > \omega_p$ and $\omega < \omega_p$.

(e) Estimate the plasma frequency ω_p for a metal and for a semiconductor with $n = 10^{16}$ cm^{-3}.

5. In a Hall effect experiment in GaAs the following values of the mobilities of the charge carriers were measured: $\mu_e = 0.850\ \text{m}^2\ \text{V}^{-1}\ \text{s}^{-1}$, $\mu_h = 0.040\ \text{m}^2\ \text{V}^{-1}\ \text{s}^{-1}$. When the semiconductor sample is illuminated by an intense light source, its conductivity increases by $\Delta\sigma = 2.95\,\Omega^{-1}\ \text{m}^{-1}$ compared to the value measured in the dark. When the light source is switched off, $\Delta\sigma$ decreases by half after 3.8×10^{-5} s. Determine the recombination time of the electron-hole pairs and the optical generation rate for the conditions of the experiment.

6. A single-crystal GaAs sample of dimensions $3 \times 2 \times 0.3\ \text{cm}^3$ is illuminated with infrared laser radiation. The absorption coefficient for the wavelength of the light beam used in the experiment ($h\nu = 1.4$ eV) is 500 cm^{-1}. The intensity of the laser beam is $5 \times 10^{-4}\ \text{W cm}^{-2}$ and the sample is at room temperature.

   (a) Calculate the photon flux incident on the semiconductor.
   (b) At what depth does the intensity of the light beam reduce to 5% of the value at the surface?
   (c) Estimate the number of electron/hole pairs created per second by the laser beam on the semiconductor.
   (d) Estimate the increase in the conductivity of GaAs due to illumination, taking the recombination time as 2×10^{-4} s ($\mu_e = 0.850\ \text{m}^2\text{V}^{-1}\text{s}^{-1}$, $\mu_h = 0.040\ \text{m}^2\text{V}^{-1}\text{s}^{-1}$).

7. A long, homogeneous bar of a doped n-type semiconductor is illuminated at one end and the stationary regime is reached. The excess concentration of minority charge carriers remains constant over time and has the value $\delta p(0) = 3.90 \times 10^{15}\ \text{m}^{-3}$. The cross-section of the bar is circular and has a radius $r = 5.00$ mm. Determine the current density corresponding to the injection of holes in the illuminated face. ($D = 8.5 \times 10^{-3}\ \text{m}^2\text{s}^{-1}$; $\tau = 4.8 \times 10^{-6}$ s)

8. *Photoconductors*

   Consider a parallelepiped semiconductor where the face with dimensions $w \times L$ is exposed to an illumination power P_{opt}, with photons of energy $E = h\nu$ larger than the bandgap. Interband absorption leads to the formation of electron-hole pairs at the generation rate:

$$G = \frac{\Delta n}{\tau} \tag{4.117}$$

   where Δn is the corresponding change in the concentration of conduction electrons and τ is the average lifetime of the charge carriers. Consider that the *quantum efficiency* (average number of electron-hole pairs generated per photon) is η. The pairs are formed in the average volume $V = w\,d\,L$, where d is the average penetration length of the photons in the semiconductor.

The semiconductor is simultaneously subjected to an electric field ε parallel to the dimension L.

(a) Show that

$$G = \eta \frac{P_{opt}/h\nu}{V} \tag{4.118}$$

where the corresponding charge generation rate per unit time («primary current») is:

$$I_p = eGV = e\eta \frac{P_{opt}}{h\nu} \tag{4.119}$$

(b) Show that the generated photocurrent $I_{ph} = J_{ph}\, w\, d = \sigma\varepsilon\, w\, d = e\mu_n \Delta n \varepsilon\, w\, d$ is

$$I_{ph} = e \left(\eta \frac{P_{opt}}{h\nu} \right) \frac{\mu_n \tau \varepsilon}{L} \tag{4.120}$$

(c) Show that the *gain* is

$$g = \frac{I_{ph}}{I_p} = \frac{v\tau}{L} = \frac{\tau}{t_r} \tag{4.121}$$

where $t_r = L/v = L/(\mu_n \varepsilon)$ is the transit time of the charge carriers through the semiconductor.

9. Consider a semiconductor optically excited at the surface (in the $x = 0$ plane), generating an excess concentration of holes $\Delta p(x = 0) = \Delta p_0$ and an excess concentration of conduction electrons $\Delta n(x = 0) = \Delta n_0$.

(a) Assume that this excess concentration is contained in a volume ΔV of depth Δx and surface σ. Calculate the charge associated with the holes in this volume.

$$\frac{\partial j_{px}}{\partial x} = -e \frac{\partial p(x)}{\partial t} \tag{4.122}$$

$$\frac{\partial j_{ex}}{\partial x} = e \frac{\partial n(x)}{\partial t} \tag{4.123}$$

(b) Assume that this charge generates a diffusion current from the surface. Determine the current density $J = \Delta q/(\sigma \Delta t)$. Define the surface recombination velocity $S = \Delta x/\Delta t$.

(c) Show that

$$S \Delta p(x = 0) = D_p \frac{\partial}{\partial x} \Delta p \bigg|_{x=0} \tag{4.124}$$

10. Obtain the following limits for eq. 4.110:

   (a) Zero recombination limit ($S = 0$)

$$\Delta p(x) = \frac{G_0 \tau_p}{1 - \alpha^2 L_p^2} \left[\exp(-\alpha x) - \alpha L_p \exp(-x/L_p)\right] \quad (4.125)$$

   (b) Fast recombination limit ($S \to \infty$)

$$\Delta p(x) = \frac{G_0 \tau_p}{1 - \alpha^2 L_p^2} \left[\exp(-\alpha x) - \exp(-x/L_p)\right] \quad (4.126)$$

11. Show that, in the fast recombination limit (eq. 4.126), the following results apply:

   (a) $\Delta p(0) = 0$

   (b) $j(x = 0) = -eD \left.\frac{\partial \Delta p}{\partial x}\right|_{x=0} = -e\frac{G_0 L_p}{1+\alpha L_p}$

   (c) $j(x = w) = e\frac{G_0 L_p}{1-\alpha^2 L_p^2} \exp(-\alpha w)\left[\alpha L_p - \exp(w\alpha - w/L_p)\right]$

   (d) Show that equation (c) can be written, in the limit $w << L_p << \alpha^{-1}$:

$$j(w) = -e\frac{G_0 L_p}{1 + \alpha L_p} \exp(-\alpha w) \quad (4.127)$$

   (e) Analyze the signs of the currents obtained in the previous paragraphs. Which direction are the currents: in the same direction as the incident light or in the opposite direction? At what depth does the diffusion current reverse direction?

12. *Recombination through impurities: Shockley-Read-Hall model*[18]

   Figure 4.4 illustrates the possible recombination processes between band levels and the energy level E_t of an impurity (here called capture center or trap). Consider that the concentration of traps is N_t and that, in thermodynamic equilibrium, the probability of the impurity level being occupied is (E_F is the position of the chemical potential/Fermi level):

$$f_t^0 = \frac{1}{1 + \exp\frac{E_t - E_F}{kT}} \quad (4.128)$$

   In the case of process (a), the capture rate r_C is proportional to the concentration of unoccupied capture centers, the concentration of

[18] Cf. W. Shockley and W. T. Read, Jr., *Statistics of the Recombinations of Holes and Electrons*, Physical Review **87**, 835 (1952); R.N. Hall, *Electron-Hole Recombination in Germanium*, Physical Review **87**, 387 (1952).

electrons in the conduction band and the (thermal) velocity v_{th} of the electrons:

$$r_C = \sigma_n v_{th} N_t (1 - f_t) n \tag{4.129}$$

σ_n is the effective capture cross-section and $v_{th} = \sqrt{3kT/m*}$. f_t and n do not necessarily correspond to values at thermodynamic equilibrium.

In the case of process (b), the emission rate is proportional to the concentration of occupied capture centers:

$$g_C = e_n N_t f_t \tag{4.130}$$

where e_n is the emission probability.

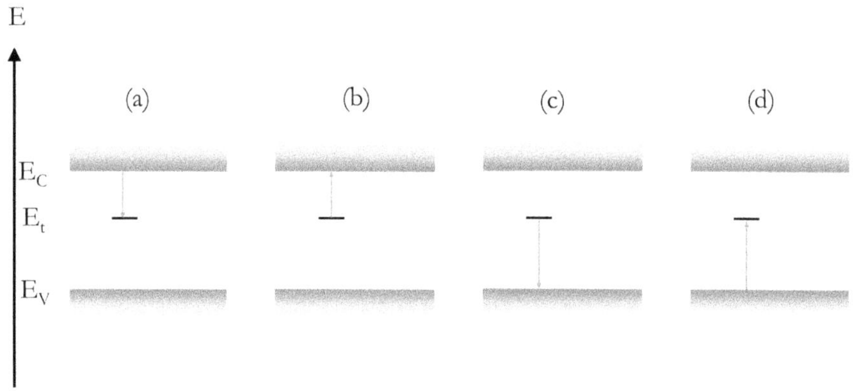

FIGURE 4.4
Schematic of the recombination processes mediated by an impurity level: (a) electron capture from the conduction band; (b) electron emission into the conduction band; (c) hole capture from the valence band; (d) hole emission into the valence band. The arrows indicate the electron transition.

(a) it is usual to use the same thermal velocity $v_{th} = \sqrt{3kT/m_0}$ for electrons and for holes, where m_0 is the mass of the free electron. Show that this corresponds to a redefinition of σ_n by a multiplicative factor.

(b) In thermodynamic equilibrium, the electron emission rate is equal to the electron capture rate. Check that it can be written:

$$e_n = \sigma_n v_{th} n_t \tag{4.131}$$

where

$$n_t = n_0 \exp \frac{E_t - E_F}{kT} \tag{4.132}$$

(c) Obtain the capture and emission rates of holes r_V and g_V, as well as the corresponding expressions for e_p and p_t.

(d) Verify that $n_t p_t = n_0 p_0$.

(e) Taking into account the conservation of electric charge ($r_C + g_V = g_C + r_V$), check that the probability of occupation of the capture centers in the case of non-equilibrium is given by:

$$f_t = \frac{\sigma_n n + \sigma_p p}{\sigma_n (n + n_t) + \sigma_p (p + p_t)} \tag{4.133}$$

(f) Show that the recombination rate r_{b-i} is:

$$r_{b-i} = \frac{np - n_0 p_0}{\tau_{p0}(n + n_t) + \tau_{n0}(p + p_t)} \tag{4.134}$$

where the lifetimes $\tau_{n0} = (\sigma_n v_{th} N_t)^{-1}$ and $\tau_{p0} = (\sigma_p v_{th} N_t)^{-1}$ were defined.

(g) Verify that the recombination rate r_{b-i} is maximum for

$$E_t = \frac{E_V + E_C}{2} + \frac{kT}{2} \ln \frac{\tau_{n0} N_V}{\tau_{p0} N_C} \tag{4.135}$$

(h) It is usual to define a lifetime $\tau = \delta n / r_{b-i}$, where δn is the excess of charge carriers in relation to the thermodynamic equilibrium situation: $n = n_0 + \delta n$, $p = p_0 + \delta n$ (where the condition of electrical neutrality in the creation of the excess carriers is assumed). Check that, if δn is negligible:

$$\tau = \frac{\tau_{p0}(n_0 + n_t) + \tau_{n0}(p_0 + p_t)}{n_0 + p_0} \tag{4.136}$$

(i) Using typical values for Si from the literature, graph $\ln \tau$ as a function of the Fermi level E_F (see figure 4.3). Identify in the figure and discuss the following cases (E_F^i represents the position of the Fermi level in the intrinsic case):

i. Strongly doped sample type p ($E_F << 2E_F^i - E_t$, $p_0 << n_t$);
ii. Weakly doped sample type p ($E_F > 2E_F^i - E_t$, $n_t >> p_0$);
iii. Weakly doped sample type n ($E_F < E_t$, $n_0 << n_t$);
iv. Strongly doped sample type n ($E_F > E_t$, $n_0 >> n_t$).

5

Semiconducting junctions

The discovery of the rectifying properties[1] of the junctions of metals with sulphides (namely PbS) was discovered by Braun[2] in 1874 and was one of the first applications of semiconductor materials. The physics of the semiconductor junction, whether bipolar (junction of semiconductors with different dopings) or unipolar (metal/semiconductor junction or Schottky junction) was only developed in the 1930s by Neville Mott and Walter Schottky[3]. The semiconductor junction remains an essential ingredient in most semiconductor applications. We start by analyzing the physics of the bipolar junction and then present some (non-exhaustive) examples of semiconductor applications.

5.1 Bipolar junction

5.1.1 Introduction

A bipolar semiconductor junction is based on the interface between two semiconductors of opposite polarities: a n type semiconductor and a p type semiconductor. The semiconductors in question can be of the same material (*homopolar* junction) or different materials (*heteropolar* junction). We start by analyzing the homopolar junction and leave for the next chapter a discussion about the case of heteropolar junctions (also often called *heterostructures*).

Consider that the junction is built from a n type semiconductor doped with a concentration N_D of donors and the same semiconductor, now doped p type with a concentration N_A of acceptors. We assume that the junction is at room temperature (the usual – but not exclusive! – operating temperature), so the concentration of electrons in the conduction band of the n semiconductor, before the junction is built, is

$$n_{n_0} \sim N_D \tag{5.1}$$

[1]The rectifying device transforms alternating current into direct current.
[2]Karl Ferdinand Braun (1850-1918).
[3]Walter Schottky (1886-1976).

DOI: 10.1201/9781003587736-5

and the concentration of holes in the valence band of the p-type semiconductor, before building the junction, is

$$p_{p0} \sim N_A \tag{5.2}$$

Figure 5.1 shows the energy levels in both semiconductors, *before* the junction is built.

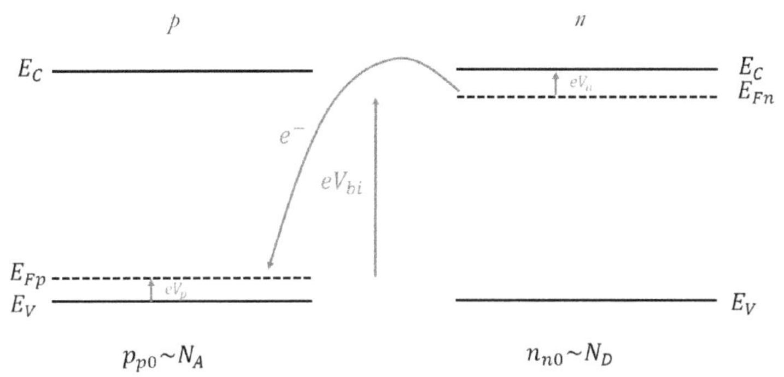

FIGURE 5.1
Energy levels in the p side and in the n side of a semiconducting junction, before the junction is built.

Since the semiconductors are the same, both the top of the valence band and the minimum of the conduction band correspond to the same energy value (E_V and E_C, respectively). In the p semiconductor the chemical potential E_{F_p} is, however, close to the top of the valence band whereas in the n semiconductor the chemical potential E_{F_n} is close to the minimum of the conduction band, as shown in the figure.

As we discussed in Chapter 1, the chemical potential is a statistical quantity that represents the change in free energy associated with the change in the number of electrons. This quantity is much higher on the n side than on the p side, so there is a considerable decrease in energy if the electrons move from the conduction band to the unoccupied levels in the valence band on the p side (remember that we are considering room temperature, so the dopants are completely ionized on both sides of the junction). During this process, charge carriers progressively disappear on both sides, thus forming a negatively charged zone on the p side and a positively charged zone on the n side (*depletion zone*). There is also the corresponding formation of an electric field in the depletion zone, which obstructs the transit of electrons from the n-type zone to the p-type zone through a potential barrier. Both are represented in figure 5.2.

We now analyze this process in more detail. As we saw in section 3.2.6.1, in the presence of an electric field the internal energy of the system depends not only on the internal degrees of freedom (associated with the chemical potential) but also on the potential energy of each particle in the presence of the electrostatic potential. The condition of thermodynamic equilibrium is thus characterized by the condition of uniformity of the electrochemical potential (Cf. eq. 3.46):

$$\nabla\left(\mu_{eq}\right)=0 \Leftrightarrow \nabla\left(\mu_q-eV\right)=0 \tag{5.3}$$

This condition allows us to redraw the level diagram of figure 5.2, adding the electrostatic potential energy to the chemical potential, thus drawing a line of constant electrochemical potential that crosses the entire junction. The redrawn diagram in figure 5.3 clearly shows the potential energy barrier eV_{bi} that the electrons have to overcome in order to cross the depletion zone, from the n zone to the p zone,[4] and simply corresponds to the difference in energy between the chemical potentials of the p zone and the n zone. This potential barrier is usually called *built-in potential*, V_{bi}.

From figure 5.1, we have:

$$\begin{aligned} eV_n &= E_C - E_{Fn} \\ eV_p &= E_{Fp} - E_V \\ E_g &= E_C - E_V \end{aligned} \tag{5.4}$$

and, as a result,

$$eV_{bi}=E_g-eV_n-eV_p \tag{5.5}$$

From the general expression 2.10, we can relate the chemical potential E_{Fn} on the n side to the concentration n_{n0} of conduction electrons on the n side, before the junction is built:

$$n_{n0}=N_C\exp\left(\frac{E_{Fn}-E_C}{kT}\right)=N_C\exp\left(-\frac{eV_n}{kT}\right) \tag{5.6}$$

Similarly, eq. 2.11 can be used to establish the relation between the chemical potential E_{Fp} on the p side and the concentration p_{p0} of holes in the valence band on the p side, before the junction is built:

$$p_{p0}=N_V\exp\left(\frac{E_V-E_{Fp}}{kT}\right)=N_C\exp\left(-\frac{eV_p}{kT}\right) \tag{5.7}$$

Taking into account eqs. 5.6 and 5.7, in connection with the law of mass action (eq. 2.12: $n_i^2=np=N_CN_V\exp(-E_g/kT)$), we can rewrite eq. 5.5 as:

$$eV_{bi}=kT\ln\frac{n_{n0}p_{p0}}{n_i^2}\sim kT\ln\frac{N_DN_A}{n_i^2} \tag{5.8}$$

[4]Note that the holes have to overcome an identical potential barrier to cross from the p zone to the n zone: the potential energy to be added in the case of holes is $+eV$.

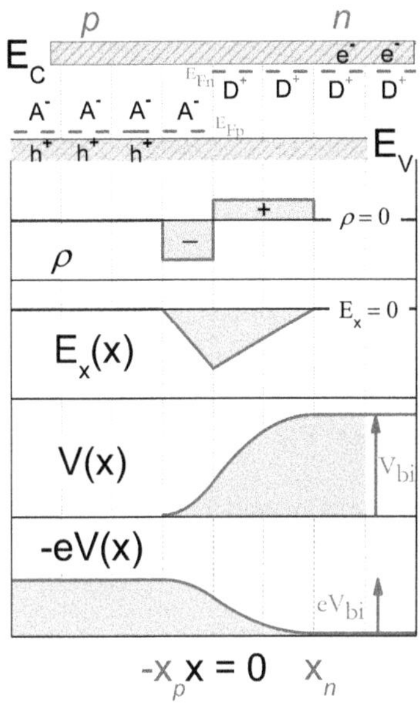

FIGURE 5.2
Electrical characterization of the depletion zone formed at the junction between a p semiconductor and a n semiconductor. From top to bottom, we show: the level diagram together with a schematic of the charge distribution in the junction, the charge density ρ in the junction, the electric field E_x, the electric potential $V(x)$ and the electric potential energy $-eV(x)$ in the depletion zone.

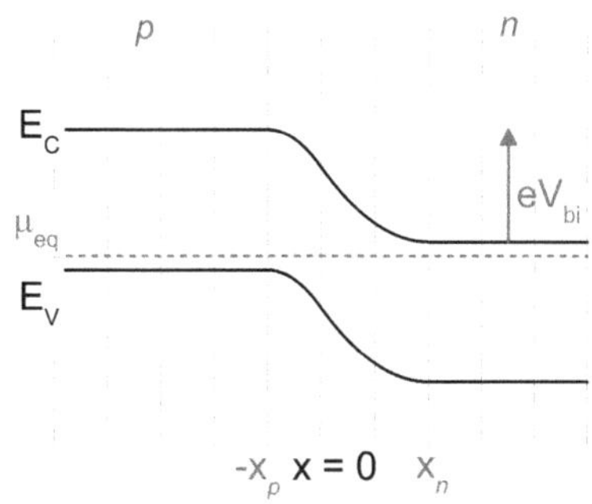

FIGURE 5.3
Level diagram of figure 5.1, redrawn by adding the electrostatic potential energy (bottom layer in Fig. 5.2) to the chemical potential, thus drawing a line of constant electrochemical potential that crosses the entire junction. This shows the potential energy barrier eV_{bi} that the electrons have to overcome to cross the depletion zone, from the n zone to the p zone.

where the usual saturation situation was taken into account: $n_{n0} \sim N_D$ and $p_{p0} \sim N_A$.

For both sides n and p, the corresponding laws of mass action (eq. 2.12) can also be established:

$$\begin{aligned} n_i^2 &= n_{n0} p_{n0} \\ n_i^2 &= n_{p0} p_{p0} \end{aligned} \tag{5.9}$$

We can thus rewrite eq. 5.8 as:

$$\begin{aligned} eV_{bi} &= kT \ln \frac{n_{n0}}{n_{p0}} \\ eV_{bi} &= kT \ln \frac{p_{p0}}{p_{n0}} \end{aligned} \tag{5.10}$$

which leads to:

$$\begin{aligned} n_{p0} &= n_{n0} \exp\left(-\frac{eV_{bi}}{kT}\right) \sim N_D \exp\left(-\frac{eV_{bi}}{kT}\right) \\ p_{n0} &= p_{p0} \exp\left(-\frac{eV_{bi}}{kT}\right) \sim N_A \exp\left(-\frac{eV_{bi}}{kT}\right) \end{aligned} \tag{5.11}$$

We therefore obtain an appreciable reduction in the concentration of charge carriers inside the depletion zone, as expected, since the dopants are neutralized in this zone.

5.1.2 Characterization of the depletion zone

We can now estimate the electric field in the depletion zone, as well as the width of this region. We start with Poisson equation[5] and assume that the donors and acceptors are completely ionized; we also assume that all conduction electrons in the n-type zone, within a width x_n from the p–n interface, flow into the p-type zone, occupying a region with width x_p. These widths are assumed to be well-defined (abrupt approximation):

$$\begin{cases} \frac{\partial^2 V}{\partial x^2} = \frac{eN_A}{\epsilon_s} \quad , \quad -x_p < x < 0 \\ \frac{\partial^2 V}{\partial x^2} = -\frac{eN_D}{\epsilon_s} \quad , \quad 0 < x < x_n \end{cases} \tag{5.12}$$

where $\epsilon_s = \epsilon_r \epsilon_0$ is the electric permittivity of the material. We thus have, for the electric field $E_x = -\frac{\partial V}{\partial x}$:

$$\begin{cases} E_x = -\frac{eN_A}{\epsilon_s} x + C_1 \quad , \quad -x_p < x < 0 \\ E_x = \frac{eN_D}{\epsilon_s} x + C_2 \quad , \quad 0 < x < x_n \end{cases} \tag{5.13}$$

Note that the depletion zone is globally neutral and that the electric field outside the depletion zone is zero. The following boundary conditions therefore apply:

$$E_x\left(x = -x_p\right) = E_x\left(x = x_n\right) = 0 \tag{5.14}$$

[5] Siméon Poisson (1781-1840).

and the integration constants C_1 and C_2 can be determined:

$$\begin{cases} E_x = -\frac{eN_A}{\epsilon_s}(x+x_p) \quad , \quad -x_p < x < 0 \\ E_x = \frac{eN_D}{\epsilon_s}(x-x_n) \quad , \quad 0 < x < x_n \end{cases} \tag{5.15}$$

The electric field is thus maximum at $x = 0$ (where the negative sign just accounts for its direction, pointing from the positively charged n-type region to the positively charged p-type region):

$$E_{max} = -\frac{eN_A}{\epsilon_s}x_p = -\frac{eN_D}{\epsilon_s}x_n \tag{5.16}$$

The continuity condition of the electric field at $x = 0$ implies:

$$N_A x_p = N_D x_n \tag{5.17}$$

This condition just informs us that the total charge on the n side is equal to the total charge on the p side (*electrical neutrality condition*). From eq. 5.15 one can also determine the electric potential across the junction (where the condition $V(x=0) = 0$ is imposed):

$$V(x) = -\int_0^x E dx = \begin{cases} \frac{eN_A}{\epsilon_s}\left(\frac{x^2}{2} + x_p x\right) \quad , \quad -x_p < x < 0 \\ -\frac{eN_D}{\epsilon_s}\left(\frac{x^2}{2} - x_n x\right) \quad , \quad 0 < x < x_n \end{cases} \tag{5.18}$$

The potential difference V_{bi} across the junction can then be calculated as (using eq. 5.16 as well):

$$\begin{aligned} V_{bi} &= V(x=x_n) - V(x=-x_p) = -E_{max}\frac{x_n}{2} - E_{max}\frac{x_p}{2} = \\ &= -\frac{1}{2}E_{max}(x_n + x_p) = -\frac{1}{2}E_{max}w \end{aligned} \tag{5.19}$$

where w is the width of the depletion layer:

$$w = x_n + x_p \tag{5.20}$$

Combining eqs. 5.16, 5.17 and 5.20 in eq. 5.19, we obtain the following:

$$V_{bi} = \frac{eN_A N_D w^2}{2\epsilon_s (N_A + N_D)} \tag{5.21}$$

The width w of the depletion layer therefore corresponds to:

$$w = \sqrt{\frac{2\epsilon_s (N_A + N_D)}{eN_A N_D} V_{bi}} \tag{5.22}$$

If the doping on one side of the junction is much larger than on the other side, the width of the junction is determined by the doping on the minority side. For example, in the case of a n^+p junction (where $N_D >> N_A$):

$$w \sim \sqrt{\frac{2\epsilon_s}{eN_A}V_{bi}} \tag{5.23}$$

5.1.2.1 Correction to the abrupt approximation

The charge distribution in the depletion zone shown in figure 5.2 is a very strict approximation and implies a discontinuous change in the concentration of electrical charge when moving from the p zone to the n zone and when moving from the depletion zone to the interior of the semiconductors (*bulk*). This approximation is designated *abrupt approximation* and is particularly poor in the transition from the p-type zone to the n-type zone. It is possible to obtain a temperature-dependent correction[6], which corresponds to an effective reduction in the voltage V_{bi}:

$$V_{bi} \rightarrow V_{bi} - \frac{2kT}{e} \tag{5.24}$$

5.1.2.2 Application of an external tension

We now consider the behavior of the junction when an external voltage V_{ext} is applied. Recall that inside the junction there is an appreciable reduction in the concentration of charge carriers, leading to an increase in the electrical resistance of the depletion zone compared to the zones outside it. When we apply a voltage to the junction, we can therefore consider that the voltage drop V_{ext} occurs essentially in the depletion zone. The width of this zone is still given by eq. 5.22, provided that the *built-in* voltage V_{bi} is replaced by the total voltage $V_{bi} \pm |V_{ext}|$.

Applying a positive voltage to the p part with respect to the n part leads to a progressive reduction in the barrier with respect to the built-in voltage V_{bi} (cf. figure 5.2), making it easier for current to pass through (*direct voltage*); while applying a negative voltage to the p part in relation to the n part leads to an increase in the barrier, making it harder for current to pass through (*reverse voltage*). The junction thus exhibits rectifying behavior, allowing the electric current to pass through when a direct voltage is applied (positive on the p part in relation to the n part). The width of the junction subjected to an external voltage V_{ext} therefore results in:[7]

$$w = \sqrt{\frac{2\epsilon_s\left(N_A+N_D\right)}{eN_AN_D}\left(V_{bi}-V_{ext}-\frac{2kT}{e}\right)} \tag{5.25}$$

---

[6]Cf. G. B. Garrett and W. H. Brattain, *Physical Theory of Semiconductor Surfaces*, Physical Review **99**, 376 (1955).

[7]The voltage is considered positive if applied in the p side and negative if applied in the n side.

We can write the dependence of the depletion layer width with applied voltage:

$$w(V_{ext}) = w_0\sqrt{1 - \frac{V_{ext}}{V_{bi} - \frac{2kT}{e}}} \sim w_0\sqrt{1 - \frac{V_{ext}}{V_{bi}}} \tag{5.26}$$

where w_0 corresponds to $w(V_{ext} = 0)$. The application of a forward (positive) voltage thus leads to a reduction in the depletion zone, whereas the application of a reverse (negative) voltage leads to an increase in the depletion zone.

5.1.2.3 Depletion zone capacity

As an external voltage is applied to the junction, the width of the depletion zone varies and this corresponds to a variation in its charge. We can thus determine the capacity of the junction:

$$C = \left|\frac{dQ}{dV}\right| \tag{5.27}$$

In order to estimate the capacity of an n^+p diode, where the doping of the n part is much larger than that of the p part, we start by taking into account the electrical neutrality condition (eq. 5.17). In this case, the width of the p part is much larger than that of the n part, so $w = x_n + x_p \sim x_p$. The charge Q accumulated in the capacitor can thus be written as $Q = eN_A w$ and the capacity of the junction then results:

$$C = \left|\frac{dQ}{dV}\right| = eN_A\frac{dw}{dV_{ext}} = \frac{\epsilon_s}{w} \tag{5.28}$$

5.1.3 *I–V* curve

We now analyze the I–V curve of the bipolar junction. We assume that the charge distribution in the depletion zone is approximately described by the abrupt approximation and that the statistics of charge carriers is adequately described by the classical Boltzmann approximation. In addition, we assume the small injection limit, where the density of minority carriers injected is much lower than the density of majority carriers. We will also assume that the generation current is zero in the depletion zone, with the conduction electron and hole currents remaining constant in this region.

Even before an external voltage is applied to the junction, some currents are already crossing the depletion zone (see figure 5.4):

- the conduction electrons on the p side (minority carriers) are accelerated by the electric field to the n side when they enter the depletion zone;
- the conduction electrons on the n side (majority carriers) diffuse to the p side due to the concentration (or chemical potential) gradient.

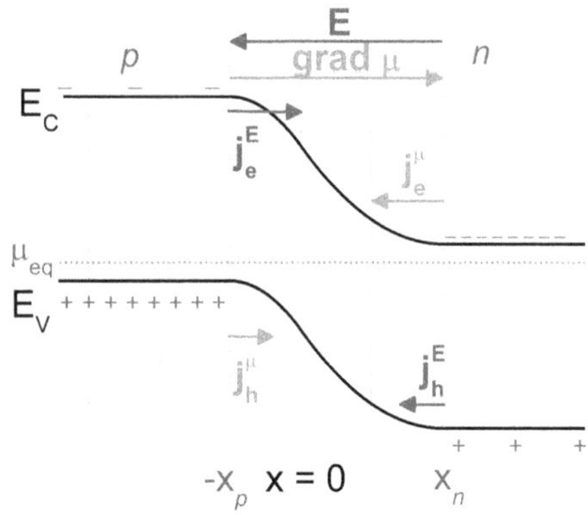

FIGURE 5.4
Representation of the different currents in the depletion layer of a $p-n$ junction, as explained in the text.

Likewise for the holes:

- the holes on the n side (minority carriers) are accelerated by the electric field to the p side when they enter the depletion zone;
- the holes on the p side (majority carriers) diffuse to the n side due to the concentration (or chemical potential) gradients.

Naturally, in thermodynamic equilibrium, the total electron and hole currents are zero. From eq. 3.51 and eq. 3.52 (where $\mathbf{E_{int}}$ is the field in the depletion zone and μ_q is the chemical potential):

$$\begin{aligned}\mathbf{J_n} &= e\,\mu_n\,n\left(\mathbf{E}_{int}+\frac{1}{e}\nabla\mu_q\right)=0\\ \mathbf{J_p} &= e\,\mu_p\,p\left(\mathbf{E}_{int}+\frac{1}{e}\nabla\mu_q\right)=0\end{aligned} \tag{5.29}$$

One aspect that needs to be emphasized is that the presence of an external voltage applied to the junction represents a situation of thermodynamic non-equilibrium, so the equilibrium conditions do not apply, namely the uniformity of the electrochemical potential across the junction. In fact, there is a difference between the «electrochemical potential» on the p side and the «electrochemical potential» on the n side, as illustrated in figure 5.5 (for the case where the applied external voltage $V_{ext} = \phi_p - \phi_n$ is negative on the p side in relation to the n side – a situation of *inverse polarization*):

$$\begin{aligned}\mu'_{eqn} &= \mu_{qn} - eV_{intn} - e\phi_n = \mu_{eqn} - e\phi_n\\ \mu'_{eqp} &= \mu_{qp} - eV_{intp} - e\phi_p = \mu_{eqp} - e\phi_p\end{aligned} \tag{5.30}$$

where μ_{qn} and μ_{qp} are the chemical potentials (in equilibrium) on the n side and on the p side, respectively; μ_{eqn} and μ_{eqp} are the respective electrochemical

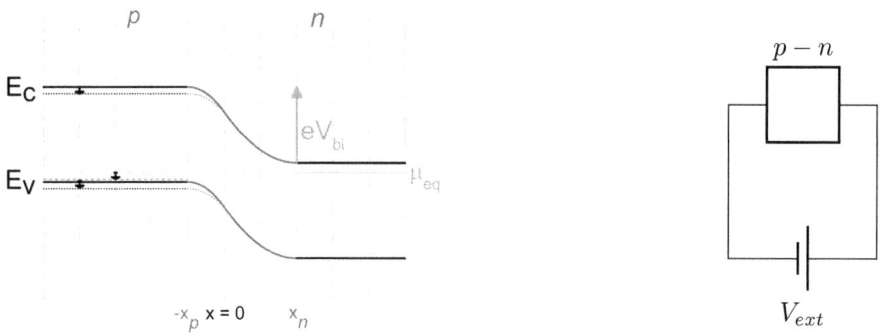

FIGURE 5.5
Modification to the diagram in figure 5.3 by applying an external voltage V_{ext} (negative on the p side in relation to the n side – *inverse polarization*). In this case, the «electrochemical potential» μ'_{eqp}, on the p side drops by the amount $-e\,V_{ext}$ in relation to the corresponding «electrochemical potential» μ'_{eqn} on the n side (Eq. 5.30).

potentials; ϕ_n and ϕ_p are the electrostatic potentials applied to the n side and to the p side, respectively.

There is thus no longer a balance between the different currents, but a total non-zero current passing through the junction.

Strictly speaking, the quantities μ'_{eqn} and μ'_{eqp} no longer represent electrochemical potentials. The corresponding quantities $\mu'_{qi} = \mu'_{eqi} + eV_{inti}$ thus also no longer correspond to chemical potentials (or Fermi levels, in the usual careless designation), and are sometimes designated *quasi Fermi levels*:

$$\begin{aligned} F_n \equiv \mu'_{qn} &= \mu'_{eqn} + eV_{intn} = \mu_{qn} - e\phi_n \\ F_p \equiv \mu'_{qp} &= \mu'_{eqp} + eV_{intp} = \mu_{qp} - e\phi_p \end{aligned} \tag{5.31}$$

The concentrations of electrons in the conduction band in the n side and of holes in the p side can be written as a function of the quasi-Fermi levels as:

$$\begin{aligned} n &= N_C \exp\left(\frac{F_n - E_C}{kT}\right) \\ p &= N_V \exp\left(-\frac{F_p - E_V}{kT}\right) \end{aligned} \tag{5.32}$$

or in function of the intrinsic concentration $n_i = p_i = N_C \exp\left((E_i - E_C)/kT\right)$:

$$\begin{aligned} n &= n_i \exp\left(\frac{F_n - E_i}{kT}\right) = n_i \exp\left[\frac{e}{kT}(\Psi - \phi_n)\right] \\ p &= p_i \exp\left(-\frac{F_p - E_i}{kT}\right) = p_i \exp\left[\frac{e}{kT}(\phi_p - \Psi)\right] \end{aligned} \tag{5.33}$$

where we have defined $E_i = -e\Psi$ and where the electrostatic potentials ϕ_n and ϕ_p can be written as :

$$\begin{aligned} \phi_n &= \Psi - \frac{kT}{e}\ln\frac{n}{n_i} \\ \phi_p &= \Psi + \frac{kT}{e}\ln\frac{p}{n_i} \end{aligned} \tag{5.34}$$

This allows us to obtain a version of the law of mass action in the non-equilibrium situation:

$$np = n_i^2 \exp\left[\frac{e}{kT}(\phi_p - \phi_n)\right] = n_i^2 \exp\left(\frac{e}{kT}V_{ext}\right) \tag{5.35}$$

Now, if a positive voltage is applied to the p side in relation to the n side (a situation known as *direct polarization*), we have

$$F_n - F_p > 0 \Rightarrow \phi_p - \phi_n > 0 \Rightarrow np > n_i^2 \tag{5.36}$$

There is therefore an increase (exponential with V_{ext}) in the product np. Note that the concentration of majority carriers remains relatively stable around the concentration of ionized dopants, so the exponential increase is mainly due to the injection of minority carriers:

$$\begin{aligned} n_p &= \frac{n_i^2}{p_p}\exp\left(\frac{eV_{ext}}{kT}\right) \\ p_n &= \frac{n_i^2}{n_n}\exp\left(\frac{eV_{ext}}{kT}\right) \end{aligned} \tag{5.37}$$

Similarly, in the situation of *inverse polarization* (negative voltage in the p side, with respect to the n side):

$$F_n - F_p < 0 \Rightarrow \phi_p - \phi_n < 0 \Rightarrow np < n_i^2 \tag{5.38}$$

This results in the suppression of minority carriers, which is also governed by eq. 5.37 (given that V_{ext} is now a negative quantity).

As we have seen, when the external voltage $V_{ext} = \phi_p - \phi_n$ is applied, a current will flow through the junction. The total currents of conduction electrons and holes passing through the depletion zone can now be written as (cf. eqs. 3.51, 3.52 and 5.29):

$$\begin{aligned} \mathbf{J_n} &= e\,\mu_n\, n\left(\mathbf{E}_{int} + \frac{1}{e}\nabla F_n\right) \\ \mathbf{J_p} &= e\,\mu_p\, p\left(\mathbf{E}_{int} + \frac{1}{e}\nabla F_p\right) \end{aligned} \tag{5.39}$$

where $\mathbf{E}_{int}$ is the electric field present in the thermodynamic equilibrium situation (the applied field due to V_{ext} is described by the quasi-Fermi levels F_n and F_p). Given eqs. 5.31 and 5.29, the currents can also be rewritten as:

$$\begin{aligned} \mathbf{J_n} &= -e\,\mu_n\, n\nabla\phi_n \\ \mathbf{J_p} &= -e\,\mu_p\, p\nabla\phi_p \end{aligned} \tag{5.40}$$

5.1.3.1 Evaluation of the current from continuity equations

The total current crossing the depletion zone can be easily evaluated from the continuity eq. 3.60, assuming that the generation rates G_n and G_p in the depletion zone are negligible:

$$\begin{aligned} D_n \nabla^2 n &= U_n = \frac{\Delta n}{\tau_n} \\ D_p \nabla^2 p &= U_p = \frac{\Delta p}{\tau_p} \end{aligned} \tag{5.41}$$

where $\Delta n = n - n_0$ and $\Delta p = p - p_0$ are the deviations from the equilibrium conduction electron and hole concentrations. Since the spatial variation of p and n is only significant in the x direction perpendicular to the junction and taking into account that $d^2n/dx^2 = d^2\Delta n/dx^2$ and $d^2p/dx^2 = d^2\Delta p/dx^2$, we obtain:

$$\begin{aligned} D_n \frac{d^2 \Delta n}{dx^2} - \frac{\Delta n}{\tau_n} &= 0 \\ D_p \frac{d^2 \Delta p}{dx^2} - \frac{\Delta p}{\tau_p} &= 0 \end{aligned} \tag{5.42}$$

As we saw earlier (eq. 5.37), the variation in the concentration of majority carriers is negligible, but the variation in the concentration of minority carriers is extremely significant. We are therefore only interested in studying the spatial variation of p in the n zone and of n in the p zone. The variation of p and n at the respective boundaries of the depletion zone can be easily estimated, given that the resistance of the depletion zone is much higher (so in practice we can consider that the external voltage V_{ext} is applied at the boundaries of the depletion zone):

$$\begin{aligned} n_p &= \frac{n_i^2}{p_p} \exp\left(\frac{eV_{ext}}{kT}\right) = n_{p0} \exp\left(\frac{eV_{ext}}{kT}\right), \quad \text{for } x = -x_p \\ p_n &= \frac{n_i^2}{n_n} \exp\left(\frac{eV_{ext}}{kT}\right) = p_{n0} \exp\left(\frac{eV_{ext}}{kT}\right), \quad \text{for } x = x_n \end{aligned} \tag{5.43}$$

The general solution of eq. 5.42 has the following form (considering that Δn is zero for $x \to -\infty$ and that Δp is zero for $x \to +\infty$):

$$\begin{aligned} \Delta n(x) &= \Delta n(x=0) \exp\left(+\frac{x}{L_n}\right) = \Delta n(x=x_0) \exp\left(+\frac{x-x_0}{L_n}\right), \quad x < -x_p \\ \Delta p(x) &= \Delta p(x=0) \exp\left(-\frac{x}{L_p}\right) = \Delta p(x=x_0) \exp\left(-\frac{x-x_0}{L_p}\right), \quad x > x_n \end{aligned} \tag{5.44}$$

where $L_n = \sqrt{D_n \tau_n}$ and $L_p = \sqrt{D_p \tau_p}$ are the diffusion lengths of conduction electrons and holes, respectively. Figure 5.6 illustrates eq. 5.44. The variations in concentration in eq. 5.44 thus result in:

$$\begin{aligned} n_p(x) - n_{p0} &= [n_p(-x_p) - n_{p0}] \exp\left(+\frac{x + x_p}{L_n}\right), \quad x < -x_p \\ p_n(x) - p_{n0} &= [p_n(x_n) - p_{n0}] \exp\left(-\frac{x - x_n}{L_p}\right), \quad x > x_n \end{aligned} \tag{5.45}$$

and, taking into account eq. 5.43:

$$\begin{aligned} n_p(x) - n_{p0} &= n_{p0} \left[\exp\left(\frac{e V_{ext}}{kT}\right) - 1\right] \exp\left(+\frac{x + x_p}{L_n}\right), \quad x < -x_p \\ p_n(x) - p_{n0} &= p_{n0} \left[\exp\left(\frac{e V_{ext}}{kT}\right) - 1\right] \exp\left(-\frac{x - x_n}{L_p}\right), \quad x > x_n \end{aligned} \tag{5.46}$$

The most significant contributions to the current in the junction come from the injection of minority carriers at the border of the depletion zone:

$$\begin{aligned} j_n(x = -x_p) &= +eD_n \left.\frac{\partial n}{\partial x}\right|_{x=-x_p} = \frac{eD_n n_{p0}}{L_n} \left[\exp\left(\frac{e V_{ext}}{kT}\right) - 1\right] \\ j_p(x = x_n) &= -eD_p \left.\frac{\partial p}{\partial x}\right|_{x=x_n} = \frac{eD_p p_{n0}}{L_p} \left[\exp\left(\frac{e V_{ext}}{kT}\right) - 1\right] \end{aligned} \tag{5.47}$$

The total current is therefore:

$$J = j_p(x = x_n) + j_n(x = -x_p) = \left(\frac{eD_p p_{n0}}{L_p} + \frac{eD_n n_{p0}}{L_n}\right) \left[\exp\left(\frac{e V_{ext}}{kT}\right) - 1\right] \tag{5.48}$$

This current is shown in figure 5.7. The curve reflects the rectifying properties of the junction, where the current in forward polarization increases exponentially with voltage but quickly saturates in reverse polarization, reaching the *saturation current*:

$$J_{sat} = \frac{eD_p p_{n0}}{L_p} + \frac{eD_n n_{p0}}{L_n} \tag{5.49}$$

In the case where one of the sides is highly doped, the saturation current will be dominated by only one of the terms. For example, in the case of a p^+n junction:

$$J_{sat} \sim \frac{eD_p p_{n0}}{L_p} = e\sqrt{\frac{D_p}{\tau_p}} \left(\frac{n_i^2}{N_D}\right) \tag{5.50}$$

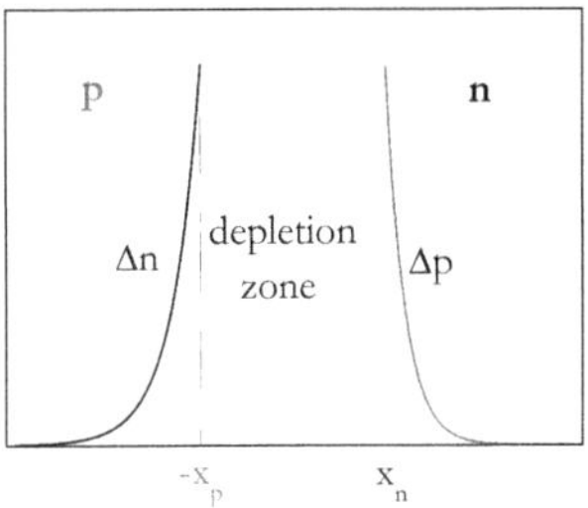

FIGURE 5.6
Injection of minority carriers by applying an external voltage, as expressed in eq. 5.44.

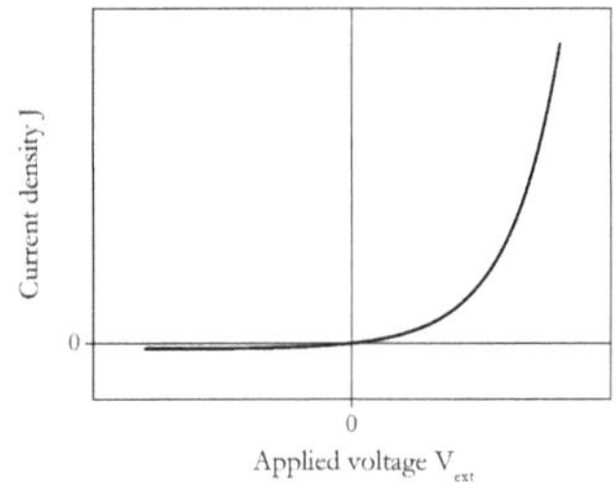

FIGURE 5.7
Sketch of the typical dependence curve of the current density passing through a $p-n$ junction subjected to an external voltage V_{ext}, at a given temperature (eq. 5.48).

5.1.3.2 Junction breakdown

In reverse polarization, the model we have just presented describes adequately the behavior of the junction for applied voltages that are not too high. As the (reverse) voltage V_{ext} increases, an abrupt increase in current can occur, known as junction breakdown. The break can be caused by three different reasons:

1. *thermal instability*: the analysis of the $p-n$ junction assumes that the intrinsic process is negligible. However, the saturation current can be affected by the local temperature increase generated by heat dissipation. Eq. 5.50 for the saturation current of a p^+n junction illustrates the effect more clearly: in eq. 5.50 it is clear that the saturation current is proportional to n_i^2, which in turn is proportional to $\exp(-E_g/2kT)$.
2. *tunnel effect*: as illustrated in figure 5.5, the application of an inverse voltage implies a reduction in energy of the position of the bands on the n side in relation to the p side. When $e(V_{bi}+V_{ext}) > eE_g$, the bottom of the conduction band on the n side becomes lower in energy than the top of the valence band on the p side, which generates an additional electron current from the n side to the p side, proportional to the transmission coefficient of the electrostatic barrier represented by the depletion zone. This effect is also known as the *Zener effect.*[8]
3. *avalanche multiplication*: Above a certain threshold, the current passing through the junction can cause impact ionization, which has an exponential current multiplication effect (an *avalanche*).

[8] Clarence Zener (1905-1993).

5.2 Diodes

A diode is a two-terminal electronic component that (ideally) conducts current only in one direction. From the analysis we have just made of the p–n semiconductor junction, it is immediately clear that the junction can be applied as a diode. An immediate application is current rectification, transforming an alternating current (whose average current is zero) into a current with a non-zero continuous component (by suppressing the negative parts of the current cycle).

The junction can also be constructed from a metal and a semiconductor (*Schottky junction*, see problem 8), giving rise to the *Schottky diode*. The Schottky diode naturally has a much lower differential resistance in direct polarization. In addition, the width of the depletion zone is smaller and, for high dopings of the semiconductor, can be so small that the transmission coefficient (due to the tunnel effect) of the charge carriers in reverse polarization increases to almost 100%, eliminating the rectifying character of the junction and thus constituting an ohmic contact.

In addition, as we have seen, junction breakdown due to the tunnel effect occurs for a well-defined voltage, which allows the junction to be used as a precise voltage reference in reverse polarization (*Zener diodes*). The *avalanche diode* has similar characteristics to the Zener diode, but the breakdown is based on the avalanche effect.

It should be noted that, as we saw in section 3.2.5, the designation Gunn diode is sometimes used for the microwave emitting device based on the Gunn effect (negative differential resistivity). This designation is misleading, as the Gunn diode is not a device with rectifying properties.

5.2.1 Photodiodes and solar cells

5.2.2 Photodiodes

Photodiodes represent a particularly important application of the semiconductor junction. The principle of operation is based on the possibility of using the intense electric field that is felt in the depletion zone of the junction to separate the electron-hole pairs generated by interband absorption.

In cases where the depletion zone is quite thin (associated with high semiconductor doping), photodiodes are particularly fast detectors. However, speed comes at the cost of conversion efficiency, as a substantial percentage of the incident photons may not be absorbed: complete absorption in the depletion zone therefore requires a wide depletion zone and/or a high absorption coefficient.

Photodiodes can operate without any voltage applied (*photovoltaic mode*). However, the detection speed can be increased by applying a reverse voltage, which increases the electric field in the depletion zone. The application

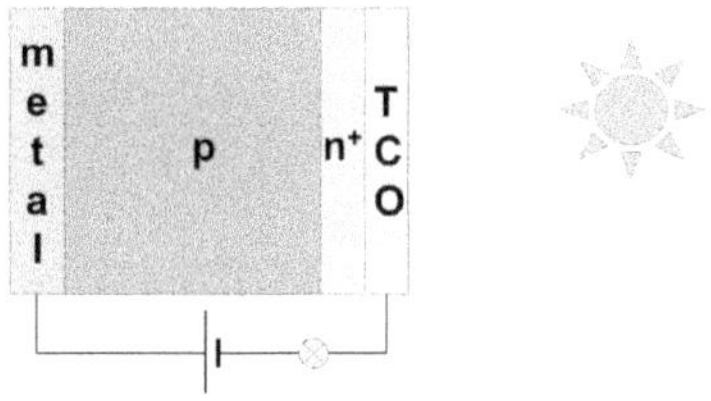

FIGURE 5.8
Schematic diagram of a solar cell based on an $n^+ - p$ semiconductor junction. The electrical contacts are typically metallic but, in the case of the external contact, the metal can be advantageously replaced by a transparent conductive oxide.

of reverse voltages close to the breakdown voltage corresponds to *avalanche photodiodes.*

5.2.3 Solar cells

The operating principle of solar cells is the same as that of photodiodes, but solar cells are optimized for maximum efficiency in obtaining electrical power from the incident light power.

The basic structure of a solar cell is schematized in figure 5.8, where an external voltage is applied to a $p - n$ junction through two conductive contacts. In order to maximize the absorption of light in the depletion zone, the upper contact (facing the sunlight) must be transparent [there has been a great deal of investment in transparent conductive oxides (TCOs)]. Likewise, the «upper» part of the semiconductor junction must be thin (as well as the corresponding depletion zone), which implies high doping. The depletion zone is thus restricted to the semiconductor «lower» part where most of the absorption of light takes place: this lower thicker part is usually designated absorber.

5.2.3.1 Current generated in ideal solar cells

The basic operating principle of a solar cell is based on the creation of electron-hole pairs in the depletion zone, which are rapidly separated by the electric field that exists there. Considering only the dominant effect of the electric field (later we will also introduce the diffusion effect), the holes are accelerated to the p zone and the electrons to the n zone, generating a photocurrent from n to p (a negative current in the usual convention used, cf. figure 5.7).

The basic effect on the I–V curve thus consists of the appearance of a negative J_L photocurrent and the corresponding shift of the I–V curve:

$$J = J_S \left[\exp\left(\frac{e\,V_{ext}}{kT} \right) - 1 \right] - J_L \qquad (5.51)$$

where the saturation current J_S is, according to eq. 5.48:

$$J_S = \frac{eD_p p_{n0}}{L_p} + \frac{eD_n n_{p0}}{L_n} \tag{5.52}$$

In this context, the saturation current corresponds to a current measured in reverse polarization even in the absence of light, and is often referred to as *dark current.*

In order to estimate the photocurrent, we consider, for simplicity, a p^+n cell, where the light absorption occurs essentially in the depletion zone of the n part (the p part corresponding to the window where sunlight comes in). The creation of electron-hole pairs by photons of higher energy than the bandgap thus creates a concentration of excess charge carriers, which decreases from the window to the end of the depletion zone. In the p^+n junction, the concentration of holes also decreases along the depletion zone from the very high value of the p zone to the low value of the n zone. Conversely, the concentration of conduction electrons increases from a low value to the value corresponding to the concentration of donors in the n zone.

Thus, the effect of the generation of electron-hole pairs from the p^+ window and along the depletion zone in the n-type region will have a much more significant impact on the reduced concentration of conduction electrons there. In order to estimate the photocurrent, we can therefore calculate the conduction electron currents. As before, we must consider the diffusion term and the drift term.

The drift current that reaches outside of the depletion zone is typically collected very quickly, the recombination along the depletion zone being negligible. The generation rate $G(x)$ at depth x is proportional to the light intensity at that depth, which varies according to the Beer-Lambert law. We therefore have:

$$G(x) = G_0 \exp(-\alpha x) \tag{5.53}$$

where, according to eq. 4.103:

$$G_0 = G_{tot}\alpha \tag{5.54}$$

The total generation rate G_{tot} simply corresponds to the total flux of photons, per unit area, transmitted inside the cells:

$$G_{tot} = (1-R)\frac{P_{inc}}{A\hbar\omega} = (1-R)\frac{P_{inc}}{Ahc}\lambda = (1-R)\,n(\lambda) \tag{5.55}$$

where R is the reflectivity of the solar cell surface, P_{inc} is the total incident optical power, A is the outer surface area of the solar cell and $\hbar\omega$ is the photon energy of angular frequency ω and wavelength λ (in vacuum). $n(\lambda)$ is the number of photons generated for a given wavelength (per unit time and per unit surface area).

Disregarding recombination in the depletion zone, the charge carriers generated there are separated by the electric field, generating a current corresponding to (for a given wavelength λ):

$$j_{drift}(\lambda) = -e\int_0^w G(x)dx = e\,G_{tot}\,[1-\exp(-\alpha w)] = \\ = e\,n(\lambda)\,[1-R(\lambda)]\left[1-\exp\left(\alpha(\lambda)w\right)\right] \tag{5.56}$$

where w is the width of the depletion zone.

The total drift current will naturally be the sum (integral) for all wavelengths:

$$j_{drift} = \int j_{drift}(\lambda)d\lambda \tag{5.57}$$

For the calculation of the contribution of diffusion to the electric current that crosses the depletion zone, we have to solve the diffusion equation, subject to the generation rate given by eq. 5.53, which leads to an expression in the form of eq. 4.110. The surface of semiconductors is usually a region of strong recombination, due to the abundance of defects and impurities. It is therefore usual to take the limit of rapid recombination at the surface (eq. 4.126, problem 10 in Chapter 3). It is also common to take the limit $w << L_p << \alpha^{-1}$ (Cf. problem 11 in Chapter 3), leading to (eq. 4.127 and eq. 5.54):[9]

$$j(w) = e\,G_{tot}\frac{\alpha L_p}{1+\alpha L_p}\exp(-\alpha w) \tag{5.58}$$

The total photocurrent is therefore (neglecting the dark current), from eqs. 5.56, 5.55 and 5.58:

$$j(w) = e\,G_{tot}\left[1-\frac{1}{1+\alpha L_p}\exp(-\alpha w)\right] = \\ = e\,(1-R)\,n(\lambda)\left[1-\frac{1}{1+\alpha L_p}\exp(-\alpha w)\right] \tag{5.59}$$

The term in square brackets in the previous equation represents the *quantum efficiency* of the solar cell, which describes the conversion efficiency of incident photons into electric current.

5.2.3.2 Engineering parameters of ideal solar cells

Figure 5.9 illustrates the typical I_V curve of a solar cell, along with some characteristic parameters of solar cells.

In reverse polarization, the junction supplies a negative current that corresponds to the photocurrent calculated earlier (eq. 5.59). In this regime, the

[9] This is now the diffusion current of conduction electrons, hence the reversal of sign.

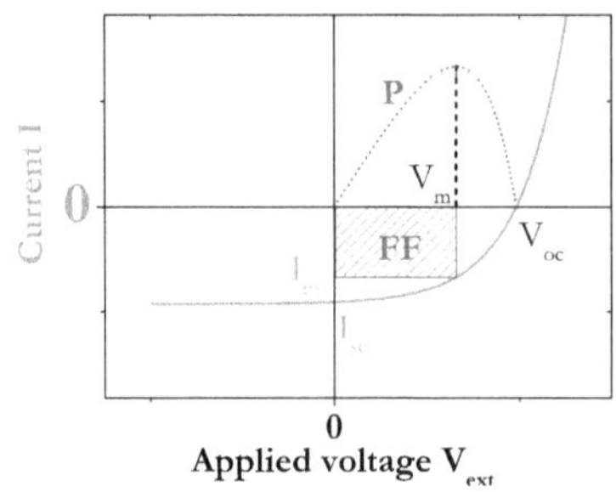

FIGURE 5.9
Sketch of the I_V curve of a solar cell, indicating the main engineering parameters, as discussed in the text.

junction basically functions as a photodetector whose current is proportional to the incident optical power. Despite this, the junction consumes the power $P = V_{ext}I$, since the external voltage supplies work to the electric charges (the current is in the same direction as the applied electric field).

However, in direct polarization and for low values of the applied external voltage, the charges move in the opposite direction of the applied external electric field, and work can be extracted from this movement. In this regime (known as the photovoltaic regime), the junction thus behaves as a generator: the solar cell.

Note that in the limit where the applied external voltage V_{ext} tends to zero (corresponding to a short circuit at the cell terminals), the current through the junction tends to the value I_{sc}, called *short-circuit current*. This current naturally corresponds to the photocurrent expressed by eq. 5.59 (for a given wavelength of the incident light).

The power $P = V_{ext}\,|I|$ generated by the solar cell is also shown in figure 5.9. For values of V_{ext} close to zero, the current through the solar cell remains relatively stable and close to the value I_{sc}, so the extracted power increases linearly with the applied voltage V_{ext}. However, when the applied voltage increases, the absolute value of the current through the junction tends to decrease as the applied electric field competes with the electric field of the depletion zone. The current thus decreases in absolute value until it completely cancels out at voltage V_{oc}. This value also corresponds to the voltage measured at the terminals of the cell in open circuit, and is thus called *open-circuit voltage*. This value establishes the operating limit at which the junction can function as a solar cell: for this voltage, the extracted power cancels out; for higher values, the junction ceases to be a generator. The voltage V_{oc} can easily be extracted from eq. 5.51:

$$V_{oc} \sim \frac{kT}{e} \ln \frac{I_L}{I_S} \tag{5.60}$$

Between these operating limits ($I = I_{sc}$ for $V_{ext} = 0$ and $I = 0$ for $V_{ext} = V_{oc}$), the extracted power increases linearly with the applied voltage V_{ext} for small voltage values, reaches a maximum $P_m = V_m I_m$ for a given voltage V_m, then decreases to zero at $V_{ext} = V_{oc}$. The voltage V_m has no analytical expression but obeys the relation:

$$V_m = V_{oc} - \frac{kT}{e} \ln\left(1 + \frac{eV_m}{kT}\right) \tag{5.61}$$

This relation can be solved graphically or numerically.

The maximum power that can be extracted from the ideal cell therefore corresponds to the maximum value $P_m = V_m I_m$ for $V_{ext} = V_m$. The pair (V_m, I_m) defines a (maximum) area enclosed on the I–V curve (marked in figure 5.9). A parameter used to assess the optimization of the cell's operation corresponds to the ratio of this area to the area defined by the pair (V_{oc}, I_{sc}) (and which circumscribes the I_V curve). This ratio is designated *Fill factor* (FF):

$$FF = \frac{V_m I_m}{V_{oc} I_{sc}} \tag{5.62}$$

5.3 Bipolar transistors

The $p-n$ junction implies the appearance of a barrier to the movement of the charge carriers, which can be manipulated by applying an external voltage. This reduces the height of the barrier in the direct direction only and favors the appearance of a current that varies exponentially with the applied voltage. This characteristic asymmetry of the junction can be explored to build a device with a set of barriers controllable by external voltages, where a directly polarized $p-n$ junction is connected to an inversely polarized $n-p$ junction. These controllable height barriers work in a way similar to a dyke that controls the flow of charge carriers between the two p parts. This device is known as a $p-n-p$ bipolar transistor[10]. The name transistor comes from the fusion of the words *transfer resistor*, since the resistances of the two junctions are generally quite different for most applications, allowing this device to be used for current amplification.

5.3.1 Minority charge carrier profiles

Figure 5.10 shows a schematic diagram of a $p-n-p$ transistor, made up of a highly doped p zone (which functions as the *emitter*), an n zone that is also highly doped (the *base*) and another p zone (the *collector*). The base-emitter junction is directly polarized with a voltage V_{EB}, while the base-collector

[10]There is also the equivalent $n-p-n$ bipolar transistor.

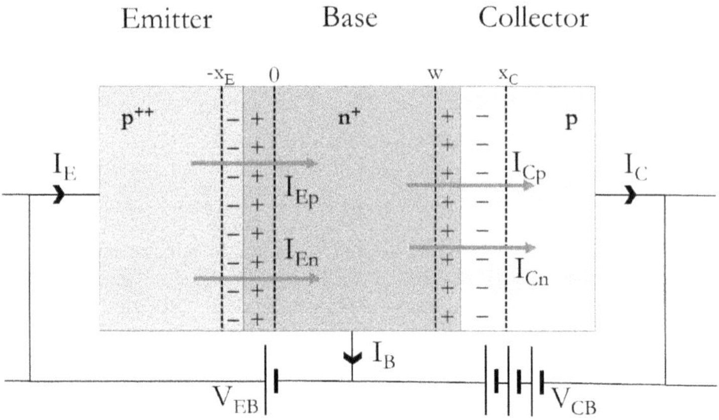

FIGURE 5.10
Schematic representation of a *p–n–p* transistor in a common base circuit, showing the depletion zones and the currents through the different parts of the system.

junction is inversely polarized with a voltage V_{CB}. The figure also shows the different currents crossing the circuit, as well as the depletion zones of the base-emitter junction (width x_E) and the base-collector junction (width $x_C - w$), separated by a neutral zone of width w, at the base. As we shall see below, for the device to work as a transistor it is essential that this width w is sufficiently small in comparison to the diffusion length of the minority carriers (in this case holes) in the base ($w << L_p$).

As we saw earlier (eq. 5.37), voltages V_{EB} and V_{CB} change significantly the concentration of minority carriers (holes in the n part and electrons in the p part). We start by analyzing the neutral zone of the base of the $p-n-p$ transistor, where the applied voltages V_{EB} and V_{CB} change the p concentration of holes in relation to the p_B equilibrium value. The profile of holes at the base can be calculated from the respective continuity equation in the steady state. Assuming a recombination rate $\Delta p/\tau_p$ (Cf. eq. 4.95), this equation reduces to:

$$\frac{d^2}{dx^2}\Delta p - \frac{1}{D_p \tau_p}\Delta p = 0 \tag{5.63}$$

where $\Delta p = p - p_B$. The general solution of this equation is, as before (eq. 4.96):

$$p(x) = p_B + C_1 \exp(x/L_B) + C_2 \exp(-x/L_B) \tag{5.64}$$

where $L_B = \sqrt{D_p \tau_p}$ is the diffusion length at the base. The integration constants C_1 and C_2 are immediately determined from the boundary conditions $\Delta p(x = 0)$ and $\Delta p(x = w)$. Now, as can be seen from eq. 5.37, the concentration of holes at the interfaces is amplified by the factor $\exp(eV/kT)$, resulting in:

$$\Delta p(x=0) = p(x=0) - p_B = p_B\left[\exp\left(\frac{eV_{EB}}{kT}\right) - 1\right] \equiv \Delta p_E \quad (5.65)$$

$$\Delta p(x=w) = p(x=w) - p_B = p_B\left[\exp\left(\frac{eV_{CB}}{kT}\right) - 1\right] \equiv \Delta p_C \quad (5.66)$$

Using the boundary conditions 5.65 and 5.66 together with the continuity eq. 5.64, we obtain the integration constants C_1 and C_2:

$$C_1 = \frac{[\Delta p(x=w) - \Delta p(x=0)\exp(-w/L_B)]}{2\sinh(w/L_B)} \quad (5.67)$$

$$C_2 = -\frac{[\Delta p(x=w) - \Delta p(x=0)\exp(w/L_B)]}{2\sinh(w/L_B)} \quad (5.68)$$

As a result, and writing $\Delta p(x=w) \equiv \Delta p_C$ and $\Delta p(x=0) \equiv \Delta p_E$:

$$p = p_B + \Delta p_E \frac{\sinh\left(\frac{w-x}{L_B}\right)}{\sinh\left(\frac{w}{L_B}\right)} + \Delta p_C \frac{\sinh\left(\frac{x}{L_B}\right)}{\sinh\left(\frac{w}{L_B}\right)} \quad (5.69)$$

An important approximation can be obtained in the case where $V_{EB} >> V_{CB}$, with the result $\Delta p_E >> \Delta p_C$ and:

$$p \sim p_B + \Delta p_E \frac{\sinh\left(\frac{w-x}{L_B}\right)}{\sinh\left(\frac{w}{L_B}\right)} \quad (5.70)$$

In the case of a wide base $w >> L_B$, we have

$$p \sim p_B + \Delta p_E \exp\left(-\frac{x}{L_B}\right) \quad (5.71)$$

In this case, the concentration of holes in the base is independent of the voltage in the collector, and the transistor does not work.

We can now proceed analogously and obtain the concentrations of conduction electrons in the p zones, characterized by the equilibrium concentrations n_E and n_C in the emitter and collector, respectively, as well as by the diffusion lengths $L_E = \sqrt{\tau_E D_E}$ and $L_C = \sqrt{\tau_C D_C}$. From the respective continuity equation:

$$\frac{d^2}{dx^2}\Delta n - \frac{1}{D_n \tau_n}\Delta n = 0 \quad (5.72)$$

we obtain the expressions

$$n(x) = n_E + \Delta n_E \exp\left(\frac{x+x_E}{L_E}\right), \text{ for } x < -x_E \quad (5.73)$$

$$n(x) = n_C + \Delta n_C \exp\left(-\frac{x-x_C}{L_C}\right), \text{ for } x > x_C \quad (5.74)$$

where we imposed the following boundary conditions:

$$\begin{aligned}\Delta n_E \equiv \Delta n(x=-x_E) = n(x=-x_E) - n_E = \\ = n_E\left[\exp\left(\frac{eV_{EB}}{kT}\right) - 1\right]\end{aligned} \tag{5.75}$$

$$\begin{aligned}\Delta n_C \equiv \Delta n(x=x_C) = n(x=c_C) - n_C = \\ = n_C\left[\exp\left(\frac{eV_{CB}}{kT}\right) - 1\right]\end{aligned} \tag{5.76}$$

as well as the trivial conditions $\Delta n(x=-\infty) = \Delta n(x=+\infty) = 0$.

5.3.2 Emitter and collector currents

We are now able to obtain the current densities in the emitter and collector from the general expression 3.41 and 3.42. The hole current densities $j_{Ep} \equiv j_p(x=0)$, at $x=0$, and $j_{Cp} \equiv j_p(x=w)$, at $x=w$, are:[11]

$$\begin{aligned}j_p(x=0) = -eD_B \left.\frac{\partial p}{\partial x}\right|_{x=0} = \\ = \frac{eD_B}{L_B}\left[\Delta p_E \coth\frac{w}{L_B} - \Delta p_C \mathrm{csch}\frac{w}{L_B}\right]\end{aligned} \tag{5.77}$$

$$\begin{aligned}j_p(x=w) = -eD_B \left.\frac{\partial p}{\partial x}\right|_{x=w} = \\ = \frac{eD_B}{L_B}\left[\Delta p_E \mathrm{csch}\frac{w}{L_B} - \Delta p_C \coth\frac{w}{L_B}\right]\end{aligned} \tag{5.78}$$

On the other hand, the current densities of the conduction electrons $j_{En} \equiv j_n(x=-x_E)$, at $x=-x_E$, and $j_{Cn} \equiv j_n(x=x_C)$, at $x=x_C$, are:

$$j_n(x=-x_E) = \frac{eD_E}{L_E}\Delta n_E \tag{5.79}$$

$$j_n(x=x_C) = \frac{eD_C}{L_C}\Delta n_C \tag{5.80}$$

The current densities in the emitter, $J_E = j_p(x=0) + j_n(x=-x_E)$, and in the colector, $J_C = j_p(x=w) + j_n(x=x_C)$, are:

$$J_E = \frac{eD_B}{L_B}\left[\Delta p_E \coth\frac{w}{L_B} - \Delta p_C \mathrm{csch}\frac{w}{L_B}\right] + \frac{eD_E}{L_E}\Delta n_E \tag{5.81}$$

$$J_C = \frac{eD_B}{L_B}\left[\Delta p_E \mathrm{csch}\frac{w}{L_B} - \Delta p_C \coth\frac{w}{L_B}\right] - \frac{eD_C}{L_C}\Delta n_C \tag{5.82}$$

[11] Remember that $\coth x = \cosh x / \sinh x$ and $\mathrm{csch}\, x = 1/\sinh x$.

5.3.3 Current amplification

The transistor is an active component, resulting from the interaction between the different currents and applied voltages in the emitter, the base and the collector, where $I_E = I_B + I_C$. Figure 5.11 shows the usual schematic representation of the $p - n - p$ transistor in electrical circuits, as well as usual circuits. In the circuit with a common base (figure 5.11 (b)), the transistor functions as a current amplifier, as we now present.

There are two contributions to the I_E current in the emitter: the I_{pE} hole current from the emitter to the base and the I_{nE} conduction electron current injected from the base:

$$I_E = I_{pE} + I_{nE} \tag{5.83}$$

Similarly, there are two contributions to the I_C current in the collector: the I_{pC} current of holes from the base to the collector and the I_{nC} current of conduction electrons from the collector to the base:

$$I_C = I_{pC} + I_{nC} \tag{5.84}$$

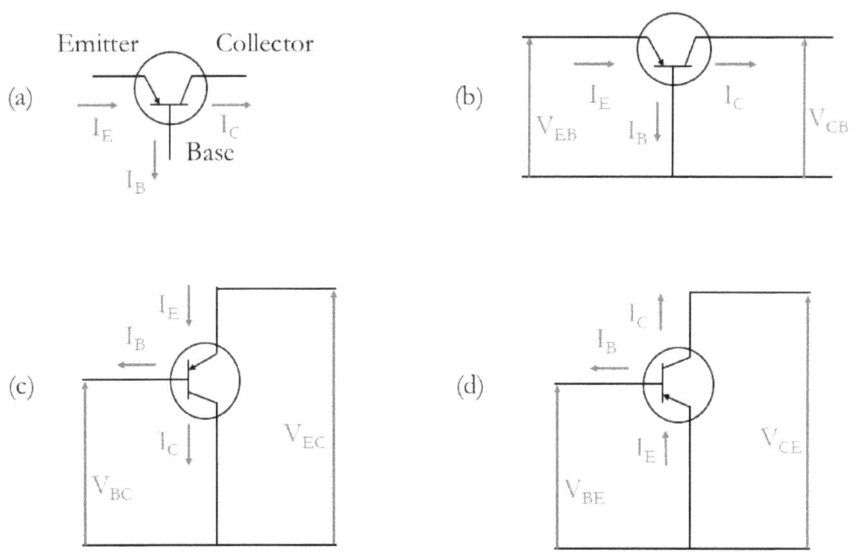

FIGURE 5.11
Usual representation of the $p - n - p$ transistor and its insertion in electrical circuits. (a) $p - n - p$ transistor, with the corresponding currents. (b) Circuit with common base. (c) Circuit with common collector. (d) Circuit with common emitter.

5.3.3.1 Circuit with common base

In the circuit with common base (figure 5.11 (b)), the current amplification (gain) is:

$$\alpha_0 = h_{FB} = \frac{\partial I_C}{\partial I_E} = \frac{\partial I_C}{\partial I_{pC}} \frac{\partial I_{pC}}{\partial I_{pE}} \frac{\partial I_{pE}}{\partial I_E} \tag{5.85}$$

and results from the product of three factors:

1. collector multiplication factor:

$$M = \frac{\partial I_C}{\partial I_{pC}}; \tag{5.86}$$

   In the usual case where the collector works below the threshold for avalanche multiplication, $M \sim 1$.

2. base transport factor α_T:

$$\alpha_T = \frac{\partial I_{pC}}{\partial I_{pE}} = \frac{j_p(x=w)}{j_p(x=0)} = \frac{\frac{\partial p}{\partial x}\mid_{x=w}}{\frac{\partial p}{\partial x}\mid_{x=0}}; \tag{5.87}$$

   Taking eq. 5.70 into account (which is applicable to the case $V_{EB} >> V_{CB}$), we get:

$$\alpha_T \sim \frac{j_p(x=w)}{j_p(x=0)} = \frac{\cosh 0}{\cosh \frac{w}{L_B}} = \frac{1}{\cosh \frac{w}{L_B}}; \tag{5.88}$$

   In the case where $w << L_B$, this result can be simplified to :

$$\alpha_T \sim 1 - \frac{w^2}{2L_B^2} \sim 1 \tag{5.89}$$

3. emitter efficiency factor γ :

$$\gamma = \frac{\partial I_{pE}}{\partial I_E} = \frac{\partial j_{pE}}{\partial J_E}\bigg|_{x=0} \sim \frac{j_{pE}}{j_{pE} + j_{nE}} = \frac{1}{1 + \frac{j_{nE}}{j_{pE}}} \tag{5.90}$$

   Taking eqs. 5.70 and 5.80 into account, we get (in the limit $V_{EB} >> V_{CB}$):

$$\frac{j_{nE}}{j_{pE}} = \frac{D_E}{D_B} \frac{L_B}{L_E} \frac{\Delta n_E}{\Delta p_E} \tanh \frac{w}{L_B} \tag{5.91}$$

   The emitter efficiency factor, in the limit $V_{EB} >> V_{CB}$, is:

$$\gamma = \left[1 + \frac{D_E}{D_B} \frac{L_B}{L_E} \frac{\Delta n_E}{\Delta p_E} \tanh \frac{w}{L_B}\right]^{-1} \tag{5.92}$$

In the case where $w << L_B, L_E$, this result can be simplified to

$$\gamma \sim \left[1+\frac{D_E}{D_B}\frac{\Delta n_E}{\Delta p_E}\frac{w}{L_E}\right]^{-1} \sim 1-\frac{D_E}{D_B}\frac{\Delta n_E}{\Delta p_E}\frac{w}{L_E} \qquad (5.93)$$

The gain α_0, in the limits $V_{EB} >> V_{CB}$ e $w << L_B, L_E$, is therefore only slightly smaller than 1:

$$\alpha_0 \sim \gamma \sim 1-\frac{D_E}{D_B}\frac{\Delta n_E}{\Delta p_E}\frac{w}{L_E} \qquad (5.94)$$

5.3.3.2 Common emitter circuit

In the common emitter circuit (figure 5.11 (d)), the gain is

$$\beta_0 = h_{FE} = \frac{\partial I_C}{\partial I_B} \qquad (5.95)$$

Since $I_E = I_B + I_C$, we have:

$$\beta_0 = h_{FE} = \frac{\partial I_C}{\partial I_B} = \frac{\partial}{\partial I_B}(I_E - I_B) = \frac{\partial I_E}{\partial I_B} - 1 = \frac{\partial I_E}{\partial I_C}\frac{\partial I_C}{\partial I_B} - 1 = \frac{\beta_0}{\alpha_0} - 1 \quad (5.96)$$

This relation can be rewritten as

$$\beta_0 = \frac{\alpha_0}{1-\alpha_0} \qquad (5.97)$$

Since the gain α_0 is only slightly smaller than 1 (eq. 5.94), the gain β_0 in the common-emitter circuit is very large. It varies from transistor to transistor, but the current amplification can reach 3 orders of magnitude.

5.3.3.3 Ebers-Moll model

Given eqs. 5.66, 5.75 and 5.76, the currents $I_E = AJ_E$ at the emitter and $I_C = AJ_C$ at the collector (from the current densities expressed in eq. 5.82, where A is the area crossed by the current) can be rewritten as follows:

$$I_E = a_{11}\left[\exp\left(\frac{eV_{EB}}{kT}-1\right)\right] + a_{12}\left[\exp\left(\frac{eV_{CB}}{kT}-1\right)\right] \qquad (5.98)$$

$$I_C = a_{21}\left[\exp\left(\frac{eV_{EB}}{kT}-1\right)\right] + a_{22}\left[\exp\left(\frac{eV_{CB}}{kT}-1\right)\right] \qquad (5.99)$$

where

$$a_{11} = eA\left[\frac{p_B D_B}{L_B}\coth\left(\frac{w}{L_B}\right) + \frac{n_E D_E}{L_E}\right] \equiv I_{ES} \qquad (5.100)$$

$$a_{12} = -a_{21} = -eA\frac{p_B D_B}{L_B}\mathrm{csch}\left(\frac{w}{L_B}\right) \qquad (5.101)$$

$$a_{22} = -eA\left[\frac{p_B D_B}{L_B}\coth\left(\frac{w}{L_B}\right) + \frac{n_C D_C}{L_C}\right] \equiv -I_{CS} \qquad (5.102)$$

I_{ES} and I_{CS} represent the junction saturation currents at the emitter and collector, respectively. We also have the «normal» gain $\alpha_N \equiv \alpha_0$ (Cf. eq. 5.85) and the «inverse» gain α_I, in the common base circuit.

$$a_{12} = \alpha_I I_{CS} \tag{5.103}$$

$$a_{21} = \alpha_N I_{ES} = -\alpha_I I_{CS} \tag{5.104}$$

In this model, the currents in the three contacts simply result in:

$$\begin{aligned} I_E &= I_{ES}\left[\exp\left(\frac{eV_{EB}}{kT}\right) - 1\right] - \alpha_I I_{CS}\left[\exp\left(\frac{eV_{CB}}{kT}\right) - 1\right] \\ I_C &= \alpha_N I_{ES}\left[\exp\left(\frac{eV_{EB}}{kT}\right) - 1\right] - I_{CS}\left[\exp\left(\frac{eV_{CB}}{kT}\right) - 1\right] \\ I_B &= (1-\alpha_N)\, I_{ES}\left[\exp\left(\frac{eV_{EB}}{kT}\right) - 1\right] + (1-\alpha_I)\, I_{CS}\left[\exp\left(\frac{eV_{CB}}{kT}\right) - 1\right] \end{aligned} \tag{5.105}$$

The basic physical behavior of the transistor can be reproduced using the set of equations in eqn. 5.105, called *Ebers-Moll equations*[12] which are governed only by three parameters I_{ES}, I_{CS} and $\alpha_N \equiv \alpha_0$.

Problems

1. Show that the typical value of w (Cf. eq. 5.23) for Si is of the order of a micrometer.
2. Combine eqs. 5.16, 5.17 and 5.20 in eq. 5.19 in order to obtain eq. 5.21 for the built-in potential.
3. Obtain the expression for the capacity of a $p-n$ junction (eq. 5.28).
4. Consider a $p-n$ junction formed on a silicon substrate at room temperature. The n side is doped with 7.80×10^{19} donors/m^3 and the doping of the p side is 4.40×10^{20} acceptors/m^3. Assume that the impurities are monovalent and completely ionized. The intrinsic concentration of charge carriers in Si is 1.50×10^{16} m^{-3} at room temperature.

   (a) Calculate the contact potential, the widths of the depletion zone in the n and p regions, and the maximum electric field in the depletion zone.

[12] Jewell Ebers (1921-1959); John Moll (1921-2011).

(b) The depletion zone acts like a capacitor, where the two regions n and p have opposite charges. Calculate the capacity per unit area of this junction.

5. The saturation current in a $p - n$ junction at room temperature is 2 fA. Sketch the graph of the characteristic $I(V)$ of the junction in the range $[-5, 1]$V. Determine the differential resistance of the junction for a forward bias of 0.25 V and for a reverse bias of 1 V. Compare the values obtained.

6. Consider the Ebers-Moll equations (eq. 5.105) used as a physical model of the bipolar junction transistor.

   (a) Determine physically reasonable values for the parameters I_{ES}, I_{CS} and $\alpha_N \equiv \alpha_0$;

   (b) Graph the collector current, I_C, as a function of the voltage V_{CB}, in a common base circuit;

   (c) Graph the collector current, I_C, as a function of the voltage V_{CE}, in a common emitter circuit.

7. Consider a p-n junction acting as a solar cell

   (a) Show that the open circuit voltage, when the junction is uniformly illuminated, is

$$V_{oc} = \frac{kT}{e} \ln\left(1 + \frac{I_{sc}}{I_i}\right)$$

   where I_{sc} is the absolute value of the short-circuit current and I_i is the saturation current of the reverse-biased junction.

   (b) Estimate V_{oc} at room temperature (300 K) for an optical generation rate of 6.20×10^{24} pairs·m^{-3}s^{-1}. The concentration of holes on the n side is 7.5×10^{14} m^{-3} and the respective diffusion length is 5.5×10^{-5} m. The concentration of electrons is 4.2×10^{14} m^{-3} on the p side and the respective diffusion length is 7.4×10^{-5} m. The recombination time of the electron/hole pairs is 9.3×10^{-7} s on both sides of the junction and the sample section is 6.4×10^{-4} m^2.

   (c) Consider that the solar cell is connected to a resistor R.

      i. Show that the power dissipated in the resistor is maximum when the potential difference across its terminals obeys the relation:

$$\left(1 + \frac{eV}{kT}\right) \exp\left(\frac{eV}{kT}\right) = 1 + \frac{I_{sc}}{I_i}$$

      ii. Determine numerically the voltage at the terminals of the solar cell for the case of maximum power at the resistor.

iii. What is the current through the junction under these conditions?
iv. What is the value of the maximum power dissipated in the resistor?

8. **Schottky diode.**

   Consider a metal-semiconductor junction. The metal has a work function $W_m = E_{vac} - E_F$, where E_{vac} is the position of the energy level corresponding to an electron infinitely far from the metal and E_F is the position of the metal chemical potential. Consider an n-type semiconductor with electron affinity χ_{sc}, where $-eV_n = E_C - E_F$, where E_C is the position of the conduction band minimum.

   (a) Draw a diagram of the band structure of the junction, before and after contact. Clearly indicate all the relevant energy levels, the effect of band-bending on the junction and the height of the barrier $F_{Bn} = W_m - \chi_{sc}$.

   (b) Verify that, after the contact is established, the potential difference V_{bi} between the metal and the semiconductor is given by

   $$V_{bi} = \frac{W_m - \chi_{sc}}{e} + V_n$$

   (c) Assuming the abrupt approximation and suitable boundary conditions, solve Poisson equation and show that the potential $\phi(x)$ and the electric field $E(x)$ in the depletion zone (of width w_0), at distance x from the contact, are given by:

   $$\begin{aligned} \phi(x) &= -V_{bi} + \frac{eN_D}{\epsilon_s}\left(w_0 x - x^2/2\right) \\ E(x) &= -\frac{eN_D}{\epsilon_s}\left(w_0 - x\right) \end{aligned}$$

   where N_D is the concentration of donors in the semiconductor and ϵ_s is the relative dielectric constant of the semiconductor.

   (d) Check that the width of the depletion zone is:

   $$w_0 = \sqrt{\frac{2\epsilon_s}{eN_D} V_{bi}}$$

   (e) Draw suitable diagrams for the band structure of the junction, with the application of an external potential V_{ext} in forward and reverse polarization. How does the width of the depletion zone change?

9. Repeat the previous problem for the case of a metal junction with a p-type semiconductor.

10. **FET**

Consider a n-type field-effect transistor (FET), illustrated in the figure below, where a current flows through a n-type semiconductor, between the *source* (S) and *drain* (D) metal contacts (subject to a voltage V_D), while applying simultaneously a voltage V_G at the *gate* (G) metal contacts. The FET is therefore a unipolar device. Assume Considering reasonable assumptions (abrupt approximation for the depletion zone, gradual approximation for the channel, adiabatic approximation for the effects of potential variation over distance):

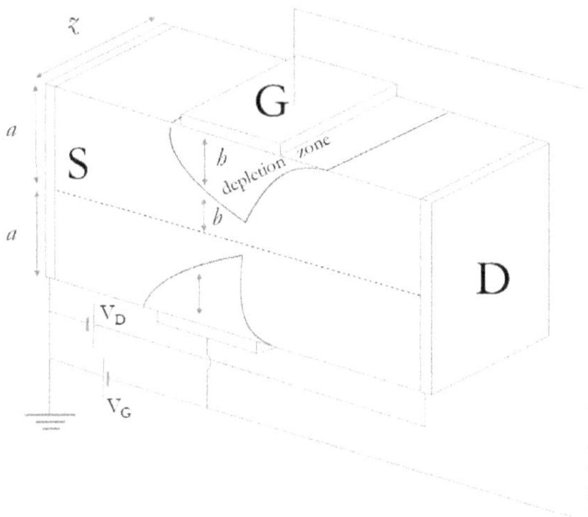

(a) Using previous calculations for the size of the depletion zone (Cf. eq. 5.26), show that the depth h of the depletion zone at position x is:

$$h = \sqrt{\frac{2\epsilon_s}{eN_D}\left(V_{bi} + V_G + V(x)\right)}$$

where $V_{bi} = \ln(N_D/n_i)/kT$ is the junction's own voltage and $V(x)$ is the source-drain voltage (*source-drain*) applied with respect to the source. Calculate the depth $y_1 = h(x = 0)$ and $y_2 = h(x = L)$.

(b) Show that the voltage $V_P = V_{bi}+V_G+V_D$ (*pinch-off tension*) for which the depth h of the depletion zone reaches its maximum value $h = a$ is:

$$V_P = \frac{eN_D a^2}{2\epsilon_s}$$

(c) Show that the drift current I_D along the channel, in the neutral zone of the semiconductor, is:

$$I_D = eN_D\mu_n \frac{\partial V}{\partial x}(a - h(x))\, z$$

(d) Consider the first law of Kirchhoff,[13] which implies that I_D cannot depend on x. Given $\int_0^L I_D dx = LI_D$ and $\partial V/\partial x = \partial V/\partial h * \partial h/\partial x$, show that:

$$I_D = I_P\left[\frac{3}{a^2}(y_2^2 - y_1^2) - \frac{2}{a^3}(y_1^3 - y_2^3)\right] =$$
$$= I_P\left[\frac{3V_D}{V_P} - 2\frac{(V_{bi}+V_G+V_D)^{3/2} - (V_{bi}+V_G)^{3/2}}{V_P^{3/2}}\right]$$

where

$$I_P = \frac{e^2 N_D^2 \mu_n z a^3}{6\epsilon_s L}$$

(e) Calculate the saturation current $I_{D,sat}$.

11. **MIS diode**

Consider an ideal metal-insulator-semiconductor (MIS) junction.

(a) Draw a diagram of the band structure of the junction, for the case of an n-type junction and for the case of a p-type junction. Clearly indicate all the relevant energy levels.

(b) Consider a junction subjected to a voltage V. For the previous situations (n-type junction and p-type junction), draw the effect of the applied voltage on the band structure, clearly distinguishing the following cases near the surface with the insulator: (i) accumulation of majority carriers; (ii) depletion of majority carriers; and (iii) reversal of the type of majority carriers.

(c) Now consider a junction where the semiconductor is p-type. Let $\psi(x)$ be the potential in the semiconductor as a function of the distance x from the insulator/semiconductor surface, $-e\psi(x) = E_i(x) - E_i(x \to \infty)$. Show that Poisson equation in the charged zone is written as:

$$\frac{d^2\psi}{dx^2} = -\frac{e}{\epsilon_s}\left[p_{p0}\left(\exp\left(-\frac{e\psi}{kT}\right) - 1\right) - n_{p0}\left(\exp\left(\frac{e\psi}{kT}\right) - 1\right)\right]$$

where p_{p0} and n_{p0} are the values of the electron and hole concentrations at *bulk* ($x = +\infty$), ϵ_s is the static dielectric constant of the semiconductor and $E_i(x)$ is the value of the intrinsic chemical potential at position x.

[13] Gustav Kirchhoff (1824-1887).

(d) Solve Poisson equation and show that the electric field in the charged region is:

$$E(x) = \pm\frac{\sqrt{2}kT}{eL_D}\left[\left[\exp\left(-\frac{e\psi}{kT}\right) - 1 + \frac{e\psi}{kT}\right] + \right.$$
$$\left. + \frac{n_{p0}}{p_{p0}}\left[\exp\left(\frac{e\psi}{kT}\right) - 1 - \frac{e\psi}{kT}\right]\right]$$

where the positive sign corresponds to the case $\psi > 0$ and the negative sign to the case $\psi < 0$. $L_D = \sqrt{\epsilon_s kT/e^2 p_{p0}}$ is the Debye[14] length. $\left(\text{N.B.}: \frac{d^2\psi}{dx^2} = \frac{1}{2}\frac{d}{d\psi}\left(\frac{d\psi}{dx}\right)^2\right)$

(e) According to Gauss law and to the boundary condition $E(x \to \infty) = 0$, show that the charge Q_s per unit area on the insulator-semiconductor surface is $Q_s = \epsilon E_s$, where E_s is the field at the surface.

(f) Use typical values for silicon from the literature and graph Q_s as a function of ψ_s, at room temperature. Indicate clearly on the graph the regions of accumulation, depletion, weak inversion and strong inversion.

[14] Peter Debye (1884-1966).

6

Nanostructures

6.1 Heterostructures and nanostructures

In the previous chapter, we analyzed the relevant phenomena to the understanding of the junction between two semiconductors. These typically occur on the spatial scale of the depletion zone, which in silicon takes on typical values of the order of a micrometer. However, the development of semiconductor growth methods has made it possible to grow multilayers of semiconductors with thicknesses of the order of a few (even one) atomic layer, of the order of a few nanometers: the corresponding semiconductor structures are called *nanostructures*. In general, the reduction in dimensionality to the order of magnitude of the nanometer can occur in one or more of the three space dimensions, leading to thin-films (nanofilms), nanotubes or quantum dots. In this chapter, where a very brief introduction to this immense topic is intended, we will focus on the reduction in one dimension only (nanofilms).

The extraordinary reduction in spatial scale often leads to unexpected phenomena and new physics. In this chapter we will also look at some of the new physical aspects associated with scale reduction.

A concept that is often associated with nanostructures, but represents a distinct question, is the case where semiconductor junctions are made up of different materials. These heterojunctions and their associations are commonly called *heterostructures*.

In heterostructures, semiconductors not only have different bandgap values, but also the energy values corresponding to the minimum of the conduction band and the maximum of the valence band are misaligned when represented on an absolute energy scale (where $E = 0$ is typically taken for electrons infinitely far from the semiconductor). On this absolute energy scale, the position of the conduction band maximum can be easily determined through the photoelectric effect, by determining the energy required to extract an electron from the semiconductor (and simultaneously mapping the valence band structure through photoelectron spectroscopy[1]). This energy value is usually called the *work function*. From this energy, the position of the minimum of the conduction band on the absolute energy scale simply results from adding

[1] Cf. Peter Y. Yu and Manuel Cardona, *Fundamentals of Semiconductors: Physics and Materials Properties*, 2nd edition, chap. 8.

DOI: 10.1201/9781003587736-6

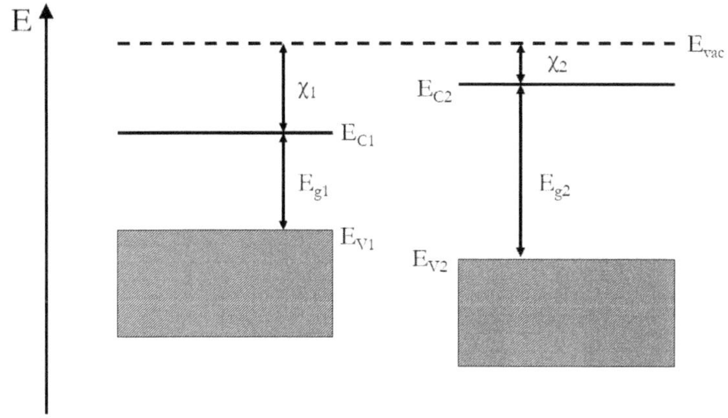

FIGURE 6.1
Level diagram of a type 1 heterostructure. The valence bands of adjacent semiconductors are shown in dashed lines and the minimum of the respective conduction bands in full lines. The position of the tops of the valence bands E_{V1} and E_{V2} and the bottoms of the conduction bands E_{C1} and E_{C2} are indicated on an absolute energy scale where E_{vac} represents the energy of the electrons not bound to the material. The electron affinities χ_1 and χ_2 are also indicated, as well as the bandgaps E_{g1} and E_{g2}.

the value of the bandgap to the energy of the maximum of the valence band. The bandgap, in turn, can be easily determined using optical spectroscopy. The resulting energy level corresponds, on the absolute energy scale, to the energy required to extract an electron from the conduction band (or the energy released when an electron in the vacuum is captured into the conduction band), and is called *electron affinity*.

6.1.1 Type I and type II heterostructures

Heterostructures can be divided into two basic types (type I and type II) depending on the relative position of the ends of the band extremes.

In type I heterostructures (figure 6.1), the energy of the electrons at the minimum of the conduction band, for the semiconductor with the lowest bandgap is lower than the energy of the electrons at the minimum of the conduction band (for the semiconductor with the highest bandgap), and the same applies to the energy of the holes at the maximum of the valence band. This leads to the band diagram shown in figure 6.1, where the semiconductor with the lowest bandgap has a higher electron affinity and lower work function than the semiconductor with the highest bandgap. In type I heterojunctions,

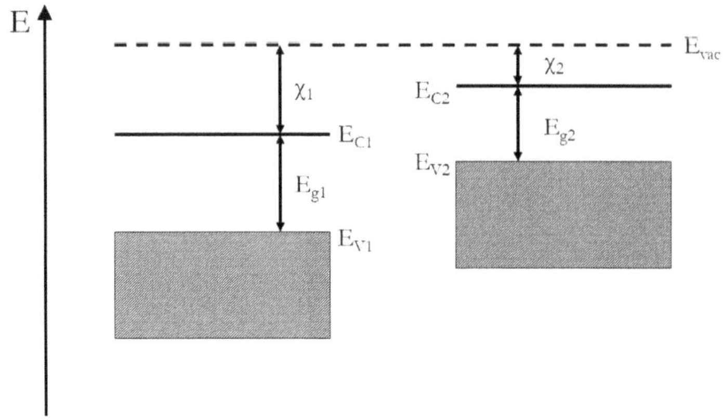

FIGURE 6.2
Level diagram of a type 2 heterostructure. The valence bands of the adjacent semiconductors are shown in dashed lines and the minimum of the respective conduction bands in full lines. The levels are indicated as in figure 6.1.

the conduction electrons will therefore tend to be located in the semiconductor with the lowest bandgap, as will the holes.

Figure 6.2 represents a type II heterostructure. Here, both the position of the minimum of the conduction band and the position of the maximum of the valence band (on the absolute energy scale) are higher for one of the semiconductors than the other. This leads to the localization of electrons in the semiconductor with the highest electron affinity and the localization of holes in the semiconductor with the lowest electron affinity.

In general, type-I heterostructures are the most relevant to the physics and technology of semiconductors.

6.1.2 Nanostructures

6.1.2.1 Growth methods

Modern methods for growing semiconductor heterostructures are essentially based on epitaxial techniques (from the greek words *epi*, meaning *over*, and *taxis*, meaning *growth*), developed in the 1960s and 1970s. The two basic methods are based on epitaxial growth from the liquid phase (LPE, *liquid phase epitaxy*) or the vapor phase (VPE, *vapor phase epitaxy*), where successive layers grow from an adjacent liquid or vapor, respectively, of controlled composition.

Since the 1980s, two new epitaxial techniques have been developed: epitaxial growth from molecular beams (MBE, *molecular beam epitaxy*) and metal-organic chemical vapor deposition (MOCVD); in MBE, growth is made from atomic or molecular beams directed to a substrate at ultra-high vacuum; in MOCVD, growth takes place in reactors in a controlled atmosphere with

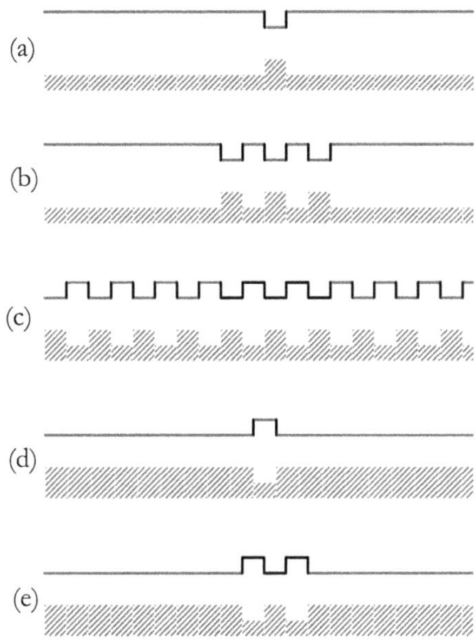

FIGURE 6.3
Examples of level diagrams of various nanostructures. The valence bands of the adjacent semiconductors are shown in dashed lines and the minimum of the respective conduction bands in full lines. The nanostructures shown correspond to: (a) quantum well; (b) multiple quantum wells; (c) super lattice; (d) single barrier tunnel structure; (e) double barrier tunnel structure.

suitable organometallic precursors, and the material is grown over a surface from the corresponding chemical reactions.

6.1.2.2 Examples

Several examples of layered structures using heterojunctions are illustrated in figure 6.3; the layers are typically a few nanometers thick, which has been made possible by the development of advanced thin-film growth techniques such as chemical vapor deposition (CVD). Therefore, these structures are generically called nanostructures and have thicknesses that enhance the overlap of electronic wave functions across the nanolayers, which favors electronic transitions between layers through the tunnel effect, as well as electronic coherence across the layers. Thus, in figure 6.3 we present:

1. quantum well structures (fig. 6.3 (a) and (b)), where one (or several) nanolayers of a semiconductor with low bandgap and high electron affinity is intercalated into a semiconductor with higher bandgap and low electron affinity, favoring the accumulation of conduction electrons and holes in this potential well;

2. structure with periodic intercalation of nanolayers of two different semiconductors (forming type I heterostructures), creating a periodic «super-potential» and giving rise to a «superlattice» (figure 6.3 (c));
3. barrier-tunnel structures (figure 6.3 (d) and (e)), where one (or several) nanolayers of a semiconductor with a high bandgap and low electron affinity is intercalated into a semiconductor with a lower bandgap and high electron affinity, creating a barrier for the movement of conduction electrons and holes.

Both quantum wells and tunneling barriers are examples of simple potentials covered in the introduction to quantum mechanics, which play a relevant role in semiconductor physics. The details of quantum wells and one-dimensional tunneling barriers can be found in any quantum mechanics textbook. We will limit ourselves here to analyzing some aspects arising from the fact that these nanolayers are in fact thin films where electrons and holes can move freely along the film, with the wells/barriers occurring only in the direction of growth.

6.2 Quantum wells

Figure 6.4 illustrates the characteristics of a model quantum well, consisting of a heterostructure of $(Al_xGa_{1-x})As/GaAs/(Al_xGa_{1-x})As$, where the disordered semiconductor $(Al_xGa_{1-x})As$ is usually abbreviated to AlGaAs, where x represents the fraction of gallium atoms replaced by aluminum atoms. The crystallographic structure of AlGaAs is the same cubic structure as GaAs and the lattice parameter is very similar, which makes these junctions particularly suitable for minimizing the tension at the interface resulting from the difference in lattice parameters. The GaAs film typically has a thickness of c. 50 nm, and the values of the bandgap and the differences between the extreme positions of the bands (band offsets) are shown in figure 6.4.

The resulting quantum well, whether for the conduction electrons in GaAs or for the holes, has finite walls (of height 0.28 eV and 0.14 eV, respectively) and the appropriate description of the corresponding energy levels can be found in any introductory quantum mechanics textbook. However, in order to obtain an approximate description, we can simply start by the model of the quantum well with infinite walls and width $W = 2a$ (figure 6.5). In this model, the wave function is forced to cancel out at the boundaries of the well (at $x = a$ and $x = -a$), resulting in the same stationary modes as a vibrating string attached at two ends, for which the possible wavelengths are:

$$\lambda_n = \frac{2W}{n} = \frac{4a}{n},\ n = 1, 2, ... \tag{6.1}$$

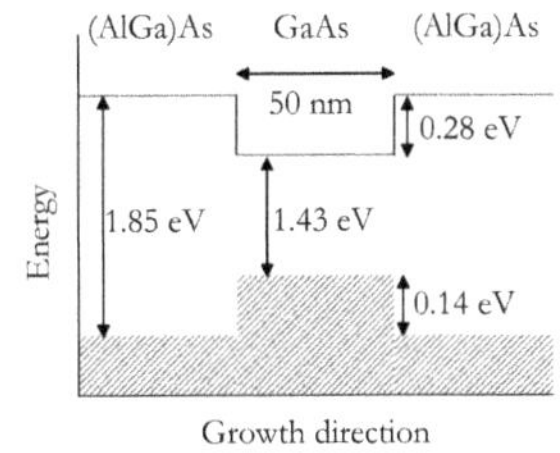

FIGURE 6.4
Level diagram of a quantum well of $(Al_{0.66}Ga_{0.34})As$ and GaAs, with an indication of the *bandgaps* and the deviations in the band positions (*band offsets*).

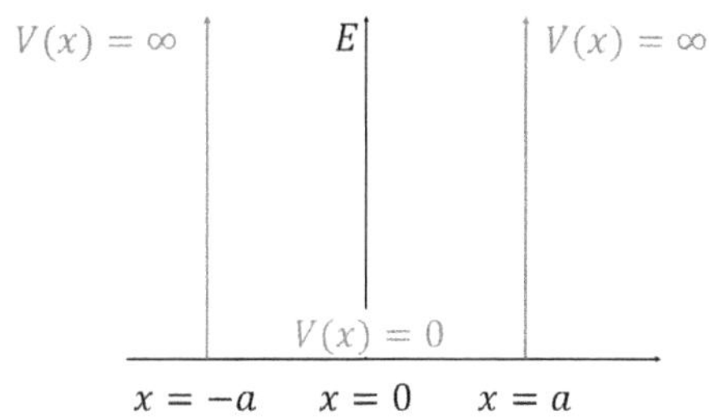

FIGURE 6.5
Model of a quantum well with infinite walls.

The corresponding linear momenta are then:

$$p_n = \hbar k_n = \frac{h}{\lambda_n} = n\frac{h}{2W}, \; n = 1, 2, ... \tag{6.2}$$

where the wavenumbers are

$$k_n = \frac{2\pi}{\lambda_n} = n\frac{\pi}{W}, \; n = 1, 2, ... \tag{6.3}$$

The energy of each mode can thus be estimated by assuming that the electron is «almost» - free in the x direction, its energy being «purely» kinetic (with the usual effective-mass correction):

$$E_n = \frac{p_n^2}{2m^*} = \frac{\hbar^2}{2m^*}k_n^2 = \frac{n^2h^2}{8m^*W^2}, \; n = 1, 2, ... \tag{6.4}$$

where the effective mass of the nanolayer has been approximated by the effective mass of the solid material (*bulk*). The corresponding wave functions, as in the vibrating string, are simply $\cos(k_n x)$ and $\sin(k_n x)$. After normalization in the interval $[x = -a, x = a]$, the normalized wave functions result in:

$$\Psi_n(x) = \begin{cases} \sqrt{\frac{2}{W}}\cos(k_n x) & , \quad n\,\text{odd} \\ \sqrt{\frac{2}{W}}\sin(k_n x) & , \quad n\,\text{even} \end{cases}$$

Note that although we did not explicitly solve the Schrödinger equation in the interval $[x = -a, x = a]$, the wave functions (eq. 6.5) and the energies (eq. 6.4) correspond to the solutions of this equation in this region:

$$-\frac{\hbar^2}{2m^*}\frac{d^2\Psi}{dx^2} = E\Psi \tag{6.5}$$

The energy levels obtained (eq. 6.4) result from the quantization of energy (and wavelengths) in the growth direction of the $(Al_xGa_{1-x})As/GaAs/(Al_xGa_{1-x})As$ films (which we are calling the x direction). However, in the plane of the film (yz plane), the potential is that of the corresponding two-dimensional GaAs lattice. For electrons at the conduction band minimum (or holes at the top of the valence band), the motion is approximately described by the parabolic approximation for quasi-free electrons and holes in the periodic potential, with the contribution to the total energy of motion in the yz plane approximately described by

$$E_{yz} = \frac{\hbar^2}{2m^*}k_y^2 + \frac{\hbar^2}{2m^*}k_z^2 \tag{6.6}$$

where $\hbar k_y$ and $\hbar k_z$ represent the components of the crystalline momentum in the y and z directions, respectively. Since the films have macroscopic dimensions in these directions, these momentum components are quasi-continuous, with the corresponding parabolic bands. The total energy of electrons and holes is therefore strongly quantized in the x direction and quasi-continuous (and approximately parabolic) in the yz plane:

$$\begin{aligned} E(n, k_y, k_z) &= \frac{n^2h^2}{8m^*W^2} + \frac{\hbar^2}{2m^*}k_y^2 + \frac{\hbar^2}{2m^*}k_z^2 = \\ &= \frac{\hbar^2}{2m^*}\left(\frac{n^2\pi^2}{W^2} + k_y^2 + k_z^2\right), \; n = 1,\, 2,\, ... \end{aligned} \tag{6.7}$$

The corresponding level diagram is schematized in figure 6.6. Each energy level associated to a given linear momentum in the direction parallel to the growth direction of the film is associated with a two-dimensional band corresponding to the momentum components perpendicular to the growth direction of the film.

As we saw in Chapter 1, the density of states associated with a two-dimensional band is constant with energy (eq. 1.26):

$$g(E) = \frac{m^*}{\pi\hbar^2} \tag{6.8}$$

In the quantum well, the density of states is simply zero for $E < E_1$ (E_1 being the energy of the ground state given by eq. 6.4). For energies between E_1 and E_2, the density of states simply corresponds to that of eq. 6.8. However, for energies above E_2, the first excited state in eq. 6.4 and the corresponding band become accessible, so the density of states suddenly doubles. The density of states function thus takes the form of a step function, as shown in figure 6.7.

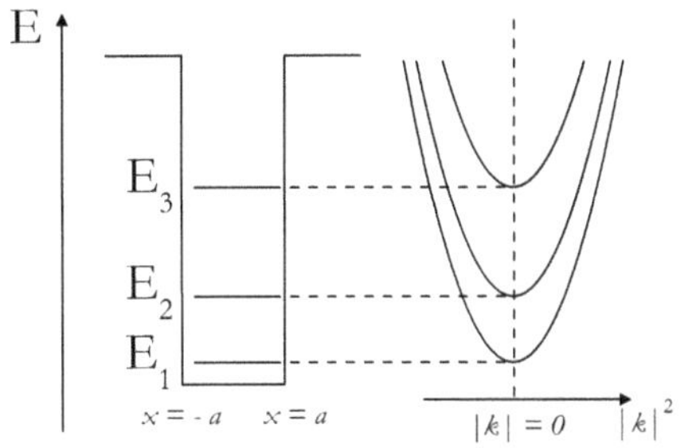

FIGURE 6.6
Energy level diagram of a quantum well: on the left, a schematic of the well in the conduction band, together with the position of the levels associated with the momentum component parallel to the direction of growth of the films; on the right, a schematic representation of the band associated with the momentum component perpendicular to the direction of growth of the films.

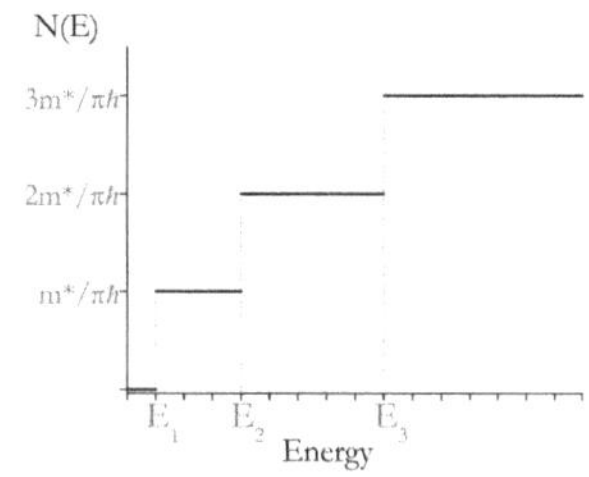

FIGURE 6.7
Density of states of a two-dimensional structure.

6.3 Superlattices

The progressive improvement and reliability of film growth methods has enabled the development of multilayer structures with periodic alternation of several tens of films with quantum well or quantum barrier characteristics. If the thickness of the barriers is small enough (compared to the wavelength of the electrons contained in the wells), the electrons can tunnel between the wells and their wave functions overlap. The periodicity associated with the alternation of the layers thus gives rise to a «super» - periodic potential in the direction of growth. As we saw in Chapter 1, energy bands are thus formed from the electronic levels of each well, with each band consisting of as many levels as the number of layers (in the order of tens).

These bands with a few tens of levels are called *minibands* and the corresponding forbidden energy intervals *minigaps*. The resulting multilayer nanostructures then constitute *superlattices*. Naturally, the width of the minibands tends to increase for smaller inter-well barrier widths because of the increasing interaction energy between the electrons in the wells, which is due to the increasing overlap of the wave functions, as illustrated in figure 6.8.

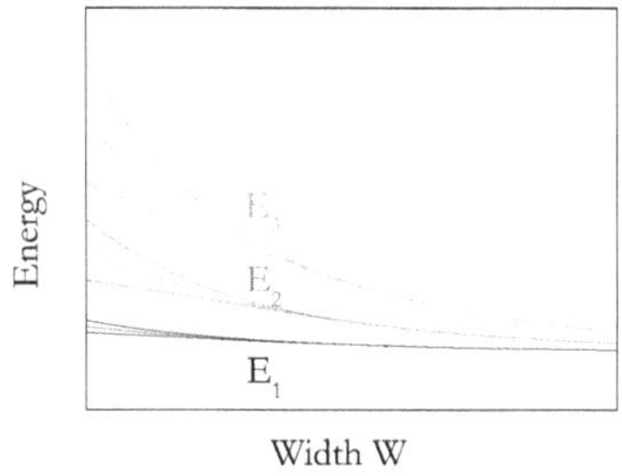

FIGURE 6.8
Schematic representation of the energy width of the mini bands of a super lattice, as a function of the width W of the quantum wells that constitute the lattice. For large W widths, the quantum wells are, in practice, isolated, and the energy levels are reduced to the levels 6.4 of a quantum well (the energy levels of an isolated quantum well are represented as dashed lines).

6.4 Quantum Hall effect

The reduced dimensionality of nanostructures prompts the appearance of effects associated with electron confinement. Among the most characteristic effects, with a major technological impact (especially in the field of metrology), are those related to the application of a magnetic field (quantum Hall effect), which we now analyze.

Remember that, for an electron in a band subject to a magnetic field **B**, the time variation of the crystalline momentum **k** is governed by the equation:

$$\hbar\left(\frac{d\mathbf{k}}{dt}\right) = -e\mathbf{v}\times\mathbf{B} \tag{6.9}$$

This equation tells us that $d\mathbf{k}/dt$ is not only perpendicular to the field **B**, but is also perpendicular to the velocity **v**, which in turn is identified with the gradient of energy with respect to **k** (eq. 1.51).

$$\mathbf{v} = \frac{1}{\hbar}\nabla_{\mathbf{k}}E(\mathbf{k}) \tag{6.10}$$

The electron's orbit thus has constant energy:

$$\frac{d\mathbf{k}}{dt}\cdot\nabla_{\mathbf{k}}E(\mathbf{k}) = 0 \Leftrightarrow \frac{\partial k_x}{\partial t}\frac{\partial E}{\partial k_x} + \frac{\partial k_y}{\partial t}\frac{\partial E}{\partial k_y} + \frac{\partial k_z}{\partial t}\frac{\partial E}{\partial k_z} = 0 \Leftrightarrow \frac{dE}{dt} = 0 \tag{6.11}$$

This is an expected result, since the magnetic field does no work.

In the **k**-space, the possible orbits of the electrons are thus described by the intersection of surfaces of constant energy with the planes perpendicular to the magnetic field **B**.

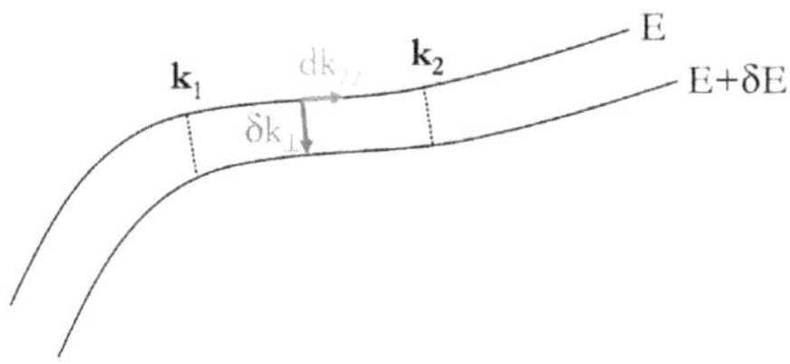

FIGURE 6.9
Interpretation of the expressions 6.15 and 6.16: adjacent orbits separated by the energy $\delta E = |\nabla_{k\perp} E|\,\delta k_\perp$, where $dk_\perp$ is the variation of **k** in the direction perpendicular the orbit (eq. 6.15).

6.4.1 Velocity components

We now consider an orbit section in **k**−space (therefore of constant energy and occurring in a plane perpendicular to the magnetic field **B**). The time interval $t_2 - t_1$ required for **k** to vary from $\mathbf{k_1}$ to $\mathbf{k_2}$ is:

$$t_2 - t_1 = \int_{t_1}^{t_2} dt = \int_{\mathbf{k_1}}^{\mathbf{k_2}} \frac{dk_{\|}}{\left|\frac{d\mathbf{k}}{dt}\right|} \tag{6.12}$$

where $dk_{\|}$ corresponds to a variation of **k** parallel to the trajectory. Considering eqs. 6.9 and 6.10, we get:

$$\frac{d\mathbf{k}}{dt} = -\frac{e}{\hbar^2}\left[\nabla_{\mathbf{k}} E(\mathbf{k})\right] \times \mathbf{B} = -\frac{e}{\hbar^2}\left|\nabla_{\mathbf{k}} E(\mathbf{k})\right| B \sin\theta = -\frac{eB}{\hbar^2}\nabla_{k\perp} E \tag{6.13}$$

where θ is the angle between $\nabla_{\mathbf{k}} E(\mathbf{k})$ and **B** and we have introduced the designation $\nabla_{k\perp} E$ for the component $\nabla_{\mathbf{k}} E(\mathbf{k})$ perpendicular to **B** (which basically corresponds to the component of the velocity perpendicular to the magnetic field).

6.4.1.1 Physical meaning of $\nabla_{k\perp} E$

In order to deepen the physical meaning of $\nabla_{k\perp} E$, recall that, from the definition of gradient, we have

$$\delta E = \nabla_{\mathbf{k}} E(\mathbf{k}) \cdot \delta\mathbf{k} \tag{6.14}$$

However, as we have seen, the possible orbits of the electrons are described by the intersection of surfaces of constant energy with planes perpendicular to the magnetic field **B**. This means that, in eq. 6.14, the product of the (non-zero) gradient $\nabla_{\mathbf{k}} E(\mathbf{k})$ with $\delta\mathbf{k}$ tells us the difference in energy δE between two different orbits separated by $\delta\mathbf{k}$. From these considerations, we have:

$$\delta E = |\nabla_{k\perp} E|\,\delta k_\perp \tag{6.15}$$

where $\delta k_\perp$ means a variation of $\mathbf{k}$ perpendicular to the trajectory. Introducing this result into eqs. 6.12 and 6.13, we obtain, for the time interval $t_2 - t_1$ needed for the value of $\mathbf{k}$ to vary, *in the same orbit*, between $\mathbf{k_1}$ and $\mathbf{k_2}$

$$t_2 - t_1 = \frac{\hbar^2}{eB\delta E} \int_{\mathbf{k_1}}^{\mathbf{k_2}} \delta k_\perp dk_\parallel \tag{6.16}$$

where $dk_\parallel$ indicates the variation in the orbit and $\delta k_\perp$ indicates the separation between adjacent orbits separated by the energy δE, as shown in figure 6.9. From this figure, it is also clear that the integral corresponds to the area A_{12} between two adjacent orbits, separated by δE, delimited by $\mathbf{k_1}$ and $\mathbf{k_2}$. We can thus write, in the limit $\delta E \to 0$:

$$t_2 - t_1 = \frac{\hbar^2}{eB} \frac{\partial A_{12}}{\partial E} \tag{6.17}$$

In the case of a closed orbit, $\mathbf{k_1} = \mathbf{k_2}$ and A_{12} represents the area bounded by the orbit (dependent on the value of the energy E and on the component of $\mathbf{k}$ parallel to the magnetic field, $k_\parallel$: $A = A(E, k_\parallel)$). The time interval $t_2 - t_1$ thus corresponds to the period of the orbit, τ_c:

$$\tau_c = \frac{\hbar^2}{eB} \frac{\partial A\left(E, k_\parallel\right)}{\partial E} \tag{6.18}$$

The respective frequency, which is nothing more than the cyclotron frequency associated with the movement (cf. eq. 3.26), is:

$$\omega_c = \frac{2\pi}{\tau_c} = \frac{eB}{m_c^*} \tag{6.19}$$

where the effective mass for cyclotron resonance is:

$$m_C^* = \frac{\hbar^2}{2\pi} \frac{\partial A\left(E, k_\parallel\right)}{\partial E} \tag{6.20}$$

As an example of the use of the above expressions, we consider a free electron subject to a magnetic field $\mathbf{B} = B\hat{e}_z$. The component of $\mathbf{k}$ parallel to $\mathbf{B}$ is thus k_z.

In $\mathbf{k}$-space, the orbit runs in the XY plane, defining an area

$$A = \pi k_{xy}^2 = \pi \left(k_x^2 + k_y^2\right) \tag{6.21}$$

The energy of the free electron is purely kinetic, so that

$$E = \frac{\hbar^2}{2m} \left(k_x^2 + k_y^2 + k_z^2\right) = \frac{\hbar^2}{2m\pi} A + \frac{\hbar^2}{2m} k_z^2 \tag{6.22}$$

The effective mass for cyclotron resonance of the free electron is therefore, from eq. 6.20:

$$m_C^* = m \tag{6.23}$$

6.4.2 Landau levels

Next, we analyze the quantum effects of applying a magnetic field to a conduction electron close to the minimum of a band. In the effective mass model, the Hamiltonian can be roughly written as

$$\frac{\hat{\mathbf{p}}^2}{2m^*}\Psi = E\Psi \tag{6.24}$$

However, remember that the Hamilton equations of classical formalism require conservative interactions $\mathbf{F}$ arising from a scalar potential U (through $\mathbf{F} = -\nabla U$), which is clearly not the case with the magnetic interaction. The Lorentz force $\mathbf{F_L} = -e(\mathbf{E} + \mathbf{v} \times \mathbf{B})$ can however be rewritten as a function of the scalar potential V and vector $\mathbf{A}$ associated with the electromagnetic field.

$$\mathbf{E} = -\nabla V - \frac{\partial \mathbf{A}}{\partial t} \tag{6.25}$$

$$\mathbf{B} = \nabla \times \mathbf{A} \tag{6.26}$$

and the following expression can be easily obtained:

$$\frac{d}{dt}\left(m\mathbf{v} + q\mathbf{A}\right) = -q\nabla\left(V - \mathbf{v}\cdot\mathbf{A}\right) \tag{6.27}$$

This expression tells us that, in the presence of a magnetic field, it is still possible to write an expression of the type

$$\frac{d\mathbf{p}_{can}}{dt} = -\nabla U(\mathbf{v}) \tag{6.28}$$

where we define the canonical momentum as:

$$\mathbf{p}_{can} = m\mathbf{v} + q\mathbf{A} \tag{6.29}$$

and the velocity-dependent potential as:

$$U(\mathbf{v}) = q\left(V - \mathbf{v}\cdot\mathbf{A}\right) \tag{6.30}$$

In the presence of a magnetic field, the Hamiltonian formalism can still be used, as long as the linear momentum is replaced by the respective canonical momentum.

The Schrödinger equation for a conduction electron in a state at the minimum of one band, in the effective mass model and in the presence of a magnetic field therefore becomes:

$$\frac{\left(\mathbf{p} - e\mathbf{A}\right)^2}{2m^*}\Psi = E\Psi \tag{6.31}$$

In the case of $\mathbf{B} = B\hat{e}_z$, a possible vector potential is (Landau gauge):[2]

$$\mathbf{A} = Bx\hat{e}_y \tag{6.32}$$

[2] Lev Landau (1908-1968).

and the Schrödinger eq. 6.31 becomes:

$$\left[\frac{p_x}{\sqrt{2m^*}}\hat{e}_x + \frac{p_y - eBx}{\sqrt{2m^*}}\hat{e}_y + \frac{p_z}{\sqrt{2m^*}}\hat{e}_z\right]^2 \Psi = E\Psi \tag{6.33}$$

Note that $[p_y, H] = [p_z, H] = 0$, so that p_y and p_z are constants of the movement and can be replaced by the respective eigenvalues $\hbar k_y$ and $\hbar k_z$:

$$\left[\frac{p_x^2}{2m^*} + \frac{(\hbar k_y - eBx)^2}{2m^*} + \frac{\hbar^2 k_z^2}{2m^*}\right]^2 \Psi = E\Psi \tag{6.34}$$

This equation can be rewritten as:

$$\left[\frac{p_x^2}{2m^*} + \frac{e^2B^2}{2m^*}\left(x - \frac{\hbar k_y}{eB}\right)\right]\Psi = \left(E - \frac{\hbar^2 k_z^2}{2m^*}\right)\Psi \tag{6.35}$$

The equation above corresponds to the equation of a harmonic oscillator of mass m^*

$$\left[\frac{p^2}{2m^*} + \frac{1}{2}m^*\omega^2 (x - x_0)^2\right]\Psi = E^*\Psi \tag{6.36}$$

where

$$x_0 = \frac{\hbar k_y}{eB} \tag{6.37}$$

$$E^* = E - \frac{\hbar^2 k_z^2}{2m^*} \tag{6.38}$$

$$\omega = \frac{eB}{m^*} \tag{6.39}$$

The eigenvalues of the Hamiltonian 6.36 are well known:

$$E^* = \left(n + \frac{1}{2}\right)\hbar\omega, \;\; n = 0, 1, 2, ... \tag{6.40}$$

Therefore, the energy levels E associated with the electron at the bottom of the conduction band and under the action of a magnetic field are (cf. eq. 6.38):

$$E = \frac{\hbar^2 k_z^2}{2m^*} + \left(n + \frac{1}{2}\right)\hbar\omega \tag{6.41}$$

$$\omega = \frac{eB}{m^*} \tag{6.42}$$

Note that these energy levels correspond to the sum of the kinetic energy component in the z direction (parallel to the field and whose corresponding velocity is unaffected by the Lorentz force) and the harmonic component corresponding to motion in the XY plane (perpendicular to the applied magnetic field, where the cyclotron orbits occur). These energies are quantized according to eq. 6.40; the $\hbar^2 k_z^2/2m^*$ component is independent of the magnetic field and quasi-continuous, with only the quantization associated with periodic boundary conditions in the z direction, resulting in a one-dimensional band with a density of states proportional to $1/\sqrt{E}$ (Cf. eq. 1.26).

The quantization of the energies corresponding to the harmonic component of the movement in the XY plane also results in the quantization of the area of the orbits in this plane perpendicular to the magnetic field. In fact, we have:

$$\left(n+\frac{1}{2}\right)\hbar\omega = \frac{\hbar^2\left(k_x^2+k_y^2\right)}{2m^*} = \frac{\hbar^2 k_{xy}^2}{2m^*} \tag{6.43}$$

The area of the orbits is therefore:

$$A = \pi k_{xy}^2 = \frac{2\pi m^*}{\hbar}\omega\left(n+\frac{1}{2}\right) \tag{6.44}$$

The successive orbits thus define «Landau tubes » in $\mathbf{k}$: cylindrical surfaces concentric with the Z axis, with a circular cross-section-space of area A. The difference δA between the areas of successive orbits is:

$$\delta A = \frac{2\pi m^*}{\hbar}\omega = \frac{2\pi eB}{\hbar} \tag{6.45}$$

6.4.3 Quantum Hall effect in two-dimensional systems

As we saw in section 6.2, a quantum well has a quantization of energy levels in the direction of growth, with quasi-continuous energy levels in the plane parallel to the direction of growth (cf. eq. 6.7), which form a two-dimensional band. By applying a magnetic field $\mathbf{B} = B\hat{e}_z$ parallel to the growth direction, the levels in this band are also quantized into Landau levels, resulting in the following set of levels:

$$E_{nl} = E_n + \left(l+\frac{1}{2}\right)\hbar\omega \tag{6.46}$$

$$E_n = \frac{\pi^2\hbar^2 n^2}{2m^* W^2} \tag{6.47}$$

In figure 6.6, the two-dimensional bands associated with each E_n level are thus replaced by a set of discrete Landau levels. The corresponding density of states would then be given by a set of Dirac functions assuming non-zero values only for the energies expressed in eq. 6.46.

However, the effect of the scattering mechanisms discussed in Chapter 3 on the movement of the electron must be taken into account. Naturally, in order for the electrons to have time to complete the cyclotron orbits, the period τ_c of the orbit must be much shorter than the average time τ between collisions:

$$\tau_c = \frac{2\pi}{\omega_c} << \tau \Rightarrow \omega_c >> \frac{1}{\tau} \tag{6.48}$$

The average time between collisions corresponds to the average lifetime of an

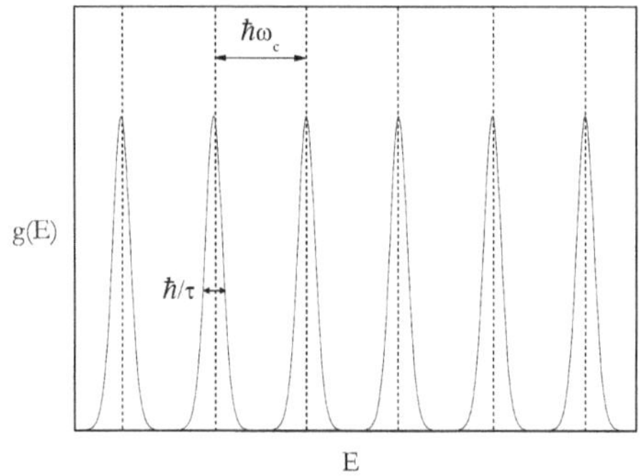

FIGURE 6.10
Schematic representation of the density of states for a subband corresponding to a given value of the quantum number n. The Landau levels, separated by the energy $\hbar\omega_c$, have a width of the order of $\hbar/\tau$, where τ is the average scattering time.

electron at a given Landau level, which is associated with the uncertainty in its energy according to the Heisenberg[3] principle:

$$\Delta E \sim \frac{\hbar}{\tau} \tag{6.49}$$

Figure 6.10 schematizes the resulting density of states for a subband corresponding to a given value of the quantum number n.

6.4.3.1 Number of electrons in each Landau level

Recall that before applying a magnetic field parallel to the z direction of growth of a quantum well, the corresponding levels can be approximately described by the free electron model, where:

$$k_x = \frac{2\pi}{L_x} n_x \tag{6.50}$$

$$k_y = \frac{2\pi}{L_y} n_y \tag{6.51}$$

The area in k-space associated with each state in the XY plane (corresponding to two electrons with opposite spins) is therefore:

$$A_1 = \frac{4\pi^2}{A_{total}} \tag{6.52}$$

where $A_{total} = L_x L_y$ corresponds to the total geometric area of the solid in the XY plane perpendicular to the magnetic field.

[3]Werner Heinsenberg (1901-1976).

The application of the magnetic field leads to the quantization of the levels in the XY plane, which «collapse» into the corresponding Landau levels. In order to estimate the number N of states associated with each Landau level, we can simply divide the area δA in k-space (eq. 6.45), corresponding to adjacent Landau levels, by the area A_1 in k-space given by eq. 6.52:

$$N = \frac{\delta A}{A_1} = \frac{2\pi eB}{\hbar}\frac{A_{total}}{4\pi^2} \tag{6.53}$$

The number of states per unit of geometric area is therefore:

$$\frac{N}{A_{total}} = \frac{eB}{h} \tag{6.54}$$

and the corresponding number N_s of electrons per unit area, in each Landau level, is:

$$N_s = \frac{2eB}{h} \tag{6.55}$$

Note, however, that in the presence of a high magnetic field, each Landau level separates into two sublevels due to the Zeeman interaction of the electron with the magnetic field, which causes a splitting in the energy of spin-up and spin-down levels:

$$\Delta E = \pm\frac{1}{2}g^*\mu_B B \tag{6.56}$$

where g^* is the effective gyromagnetic factor and μ_B the Bohr magneton. This implies that each sublevel has only half the degeneracy, leading to:

$$N_s = \frac{eB}{h} \tag{6.57}$$

6.4.3.2 Shubnikov-de Haas oscillations

This analysis allows us to conclude that the behavior of the electron gas confined in a quantum well and subject to a magnetic field parallel to the direction of growth oscillates, at low temperatures, between two extreme behaviors, illustrated in figure 6.11:

- if the chemical potential is between two Landau levels (figure 6.11 (right)), the lower energy level is completely filled, whereas the higher energy level is empty. The number of electrons per unit area is therefore (ignoring spin effects) a multiple of $2eB/h$:

$$N_s = m\frac{2eB}{h}, \quad m \text{ integer}; \tag{6.58}$$

  We thus have a case similar to that of an insulator (or semiconductor at low temperature) with low electrical conductivity;

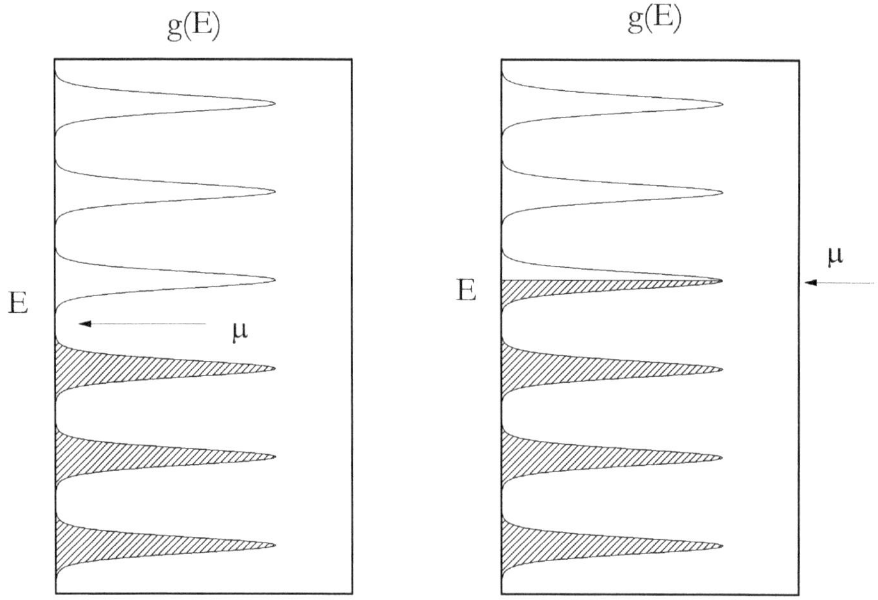

FIGURE 6.11
Extreme cases of the occupation of Landau levels: **Left** – highest energy level completely occupied, leading to low electrical conductivity, «insulator-type»; **Right** – highest energy level half-filled, leading to higher electrical conductivity, «metal-type».

- if the chemical potential is inside a Landau level (figure 6.11 (left)), this level is half-filled. In the extreme case where the level is 50% occupied, the number of electrons per unit area is:

$$N_s = \left(m + \frac{1}{2}\right)\frac{2eB}{h}, \quad m \text{ integer}; \tag{6.59}$$

We now have a case similar to that of a metal, with a higher conductivity.

The variation in the magnetic field thus gives rise to variations in electrical conductivity, obtaining minimum values for

$$\frac{1}{B} = m\frac{2e}{hN_s}, \quad m \text{ integer} \tag{6.60}$$

and maximum values for

$$\frac{1}{B} = \left(m + \frac{1}{2}\right)\frac{2e}{hN_s}, \quad m \text{ integer} \tag{6.61}$$

Electrical magnetoconductivity is therefore periodic with $1/B$, with a period of

$$\Delta\left(\frac{1}{B}\right) = \frac{2e}{hN_s} \tag{6.62}$$

Oscillations in magnetoconductivity are called Shubnikov-de Haas oscillations.[4]

6.4.3.3 The effect on resistivity

The magnetoconductivity tensor in the two-dimensional case can be written from eq. 3.33, as well as the corresponding resistivity tensor (from eq. 3.27):

$$\sigma = \frac{\sigma_0}{1+(\omega_c\tau)^2}\begin{pmatrix} 1 & -\omega_c\tau \\ \omega_c\tau & 1 \end{pmatrix} \tag{6.63}$$

$$\rho = \sigma^{-1} = \frac{1}{\sigma_0}\begin{pmatrix} 1 & \omega_c\tau \\ -\omega_c\tau & 1 \end{pmatrix} \tag{6.64}$$

σ_0 represents, as before, the value of static conductivity, with the necessary adaptation to the two-dimensional nature we are now analyzing. In this case, we now have:

$$\mathbf{J} = -N_s e\mathbf{v} \tag{6.65}$$

where the current density $\mathbf{J}$ corresponds to the current per unit length and N_s is the corresponding charge density (charge per unit area), thus resulting in:

$$\sigma_0 = \frac{N_s e^2\tau}{m^*} \tag{6.66}$$

As analyzed in section 3.2.3, for the relation between the field E_x applied longitudinally and the Hall field E_y (Cf. figure 3.2 and eq. 3.29):

$$E_y = -\omega_c\tau E_x \tag{6.67}$$

It also results, for the Hall coefficient (eq. 3.30), given that $J_x = -eN_s v_x = \sigma_{xx}E_x + \sigma_{xy}E_y$:

$$R_H = \frac{E_y}{J_x B} = \frac{\rho_{yx}}{B} = -\frac{1}{N_s e} \tag{6.68}$$

As before, the density of charge carriers can be obtained, in the low-field limit ($\omega_c\tau \to 0$), through the Hall coefficient R_H. In the high field limit, we obtain:

$$\rho_{xx} = \frac{1}{\sigma_0} = \frac{B}{N_s e}\frac{1}{\omega_c\tau} \to 0 \tag{6.69}$$

$$\rho_{yx} = \frac{-\omega_c\tau}{\sigma_0} = -\frac{B}{N_s e} = R_H B \tag{6.70}$$

and, for the corresponding components of the conductivity tensor:

[4] Lev Shubnikov (1901-1937); Wander de Haas (1878-1960).

$$\sigma_{xx} = \frac{\sigma_0}{(\omega_c \tau)^2} = \frac{N_s e}{B} \frac{1}{\omega_c \tau} \to 0 \tag{6.71}$$

$$\sigma_{yx} = \frac{\sigma_0}{\omega_c \tau} = \frac{N_s e}{B} \tag{6.72}$$

An important detail in the high-field regime is the following:

$$\rho_{xx} \propto \sigma_{xx} \propto \frac{1}{\omega_c \tau} \to 0 \tag{6.73}$$

This detail is a consequence of the different way where the conductivity and resistivity tensors are defined ($\mathbf{J} = \sigma \mathbf{E}$ and $\mathbf{E} = \rho \mathbf{J}$). Thus, $\sigma_{xx} = 0$ means no current J_x due to the application of a longitudinal field E_x, which is a consequence of the electrons performing cyclotron orbits that cancel out the average velocity in the x direction (and therefore $J_x = -eN_s < v_x >= 0$). On the other hand, $\rho_{xx} = 0$ means that the presence of a current component J_x leads to $E_x = 0$ and therefore to a drop in electric potential which is practically zero, implying a zero resistance as well, since the scattering phenomena in the x direction are in fact negligible.

Figure 6.12 shows experimental results for ρ_{xx} and ρ_{xy} as a function of magnetic field. For low fields (below 10 kG $\equiv$ 1 T), ρ_{xy} is approximately linear with B, while ρ_{xx} decreases from the value at zero field. However, for high field values, ρ_{xy} has very well-defined plateaus corresponding to values:

$$\rho_{xy} = \frac{h}{je^2}, \text{ j integer} \tag{6.74}$$

These values can be qualitatively explained using eq. 6.70 and considering that for these field values we have to include the effect of the Zeeman splitting of the Landau levels. This leads to $N_s = jeB/h$, where j is the number of Landau levels occupied, so:

$$\rho_{xy} = \frac{B}{N_s e} = \frac{h}{je^2} = \frac{25812.8\,\Omega}{j} \tag{6.75}$$

The value of these thresholds thus depends only on the fundamental constants h and e, with $h/e^2 = 25812.8\,\Omega$ being the *constant of von Klitzing.*[5] This constant plays a fundamental role in metrology: the quantum Hall effect provides an extremely precise resistance standard.

Problems

1. Consider a type-II heterostructure. The creation, by optical excitation, of an electron-hole pair in the semiconductor with the lower

[5]Klaus von Klitzing (b. 1943).

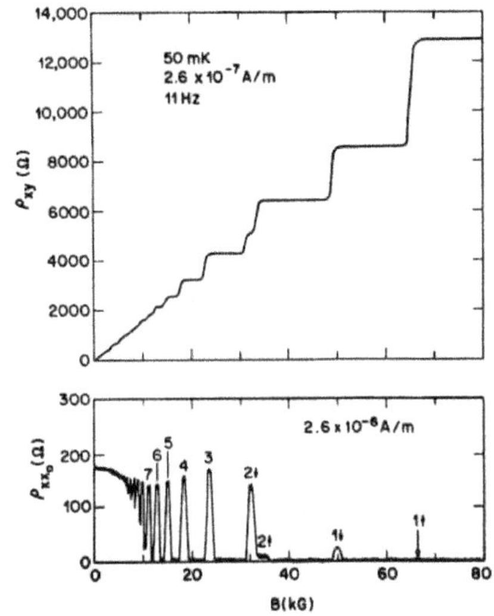

FIGURE 6.12
ρ_{xx} and ρ_{yx} as a function of the magnetic field B for a GaAs/AlGaAs heterostructure (Figure reproduced with permission from M. A. Paalanen, D. C. Tsui and A. C. Gossard, *Quantized Hall effect at low temperatures*, Phys. Rev. B **25**, 5566 (1982). *Copyright* (1982) of the American Physical Society).

work function, may give rise to an interlayer exciton, where one of the elements of the pair moves to the other semiconductor. Indicate where each member of the interlayer exciton is located and estimate the energy levels of the exciton.

2. Consider the quantum well of figure 6.4.
   (a) Assume the approximation of the quantum well with infinite walls and use eq. 6.4 to estimate the energies of the first three states in the conduction band. Consider that the effective mass of electrons in GaAs is $0.067m_0$ (with m_0 the mass of the electron).
   (b) Is this an adequate approximation?
3. Consider figure 6.12. Estimate the density of charge carriers N_s.

Bibliography

The physics of semiconductors and nanostructures is a fundamental subject in physics and engineering, so there is an appreciable amount of high-quality scientific and introductory texts available, both in English and Portuguese. This bibliography is not intended to be exhaustive, but merely a significant set of reference works that have been particularly consulted in the preparation of this work. The works that served as the main bibliography are indicated with an asterisk (*), and a summary is also provided.

* ASHCROFT, N. W. and MERMIN, N. D. – *Solid State Physics*, Fort Worth: Harcourt College Publishers, 1976. ISBN 0030839939.

This classic introduction to solid state physics (including semiconductor materials) remains a reliable and indispensable reference work.

* BARDEEN, J. – *Surface States and Rectification at a Metal Semi-Conductor Contact*, Physical Review **71**, 717 (1947).

This is a classic article by John Bardeen (1908-1991; the only physicist to have been twice awarded the Nobel Prize for Physics), where the semiconductor junction is analyzed in depth. It is possibly one of the most influential articles in semiconductor physics, although the secondary bibliography (presumably more pedagogic) fails to reproduce the clarity of the original. Essential reading.

BLAKEMORE, J. S. – *Semiconducting and other major properties of gallium arsenide*, Journal of Applied Physics **53**, R123 (1982).

BLAKEMORE, J. S. – *Semiconductor Statistics*, New York: Dover Publications, Inc., 1987. ISBN 0486653625.

BRANSDEN, B. H. and JOACHAIN, C. J. – *Physics of Atoms and Molecules*, 2nd edition. Harlow: Pearson Education Limited, 2003. ISBN 058235692X.

CHEN, S., YANG, J.H., GONG, X. G., WALSH, A. and WEI, S.-H., *Intrinsic point defects and complexes in the quaternary kesterite semiconductor* Cu_2ZnSnS_4, Physical Review B **81**, 245204 (2010).

DESNICA, U. V., DESNICA-FRANKOVIC, I. D., MAGERLE, R. and DEICHER, M., *Compensating defects and electrical activation of donors in CdS*, Physica B **273-274**, 907 (1999).

* FOX, M. – *Optical properties of solids*, 2nd edition. Oxford: Oxford University Press, 2010. ISBN: 0199573379.

This book, from the Oxford Master Series in Condensed Matter Physics, was the main bibliographical source for the chapter dedicated to optical properties, given its introductory and very pedagogical nature.
GARRETT, C. G. B. e BRATTAIN, W. H. – *Physical Theory of Semiconductor Surfaces*, Physical Review **99**, 376 (1955).

GÖTZ, W., JOHNSON, N.M., WALKER, J., BOUR, D. P. and STREET, R. A. – *Activation of acceptors in Mg-doped GaN grown by metalorganic chemical vapor deposition*, Applied Physics Letters **68**, 667 (1996).

* GRUNDMANN, M. – *The Physics of Semiconductors: an introduction including nano physics and applications*, 2nd edition. Berlin: Springer, 2010. ISBN 3642138837.

This is a contemporary and extremely comprehensive (almost encyclopedic) work, presenting semiconductor physics from the most basic aspects to the details of applications, based on a primary bibliography that reaches almost 1500 entries. This book is the basis for a two-semester Semiconductor Physics course at the University of Leipzig.

JACKSON, J. D. – *Classical eletrodynamics*, 3rd edition, New York: John Wiley & Sons, Inc., 1999. ISBN 9780471309321.

* KITTEL, C. – *Introduction to Solid State Physics*, 7th edition. New York: John Wiley & Sons, 1996. ISBN 0471111813.

This is a classic work on solid-state physics, which remains an indispensable introductory reference. The different subjects are always presented using the deep physical intuition of the author (one of the most influential theoretical physicists in solid-state physics in the second half of the 20th century) and with a permanent connection to experimental results. It is particularly recommended as a complementary bibliography for the introductory chapter on band theory.

LYSSENKO, V. G., VALUSIS, G., LÖSER, F., HASCHE, T. and LEO, K. - *Direct Measurement of the Spatial Displacement of Bloch-Oscillating Electrons in Semiconductor Superlattices*, Physical Review Letters **79**, 301 (1997).

MOMMA, K. and IZUMI, F. - *VESTA 3 for three-dimensional visualization of crystal, volumetric and morphology data*, Journal of Applied Crystallography **44**, 1272-1276 (2011).

* MISHRA, U. K. and SINGH, J. – *Semiconductor Device Physics and Design.* Dordrecht: Springer, 2008. ISBN 9781402064807.

This book provides a particularly clear overview of semiconductor applications and is therefore recommended as a complementary bibliography for this topic.

NAKAMURA, S. – *Nobel Lecture: Background story of the invention of efficient blue InGaN light emitting diodes*, Reviews of Modern Physics **87**, 1139 (2015).

NEAL, A. T., MOU, S., RAFIQUE, S., ZHAO, H., AHMADI, E., SPECK, J. S., STEVENS, K. T., BLEVINS, J. D., THOMSON, D. B., MOSER, N., CHABAK, K. D. and JESSEN, G. H. – *Donors and deep acceptors in* $\beta - Ga_2O_3$, Applied Physics Letters **113**, 062101 (2018).

PAALANEN, M. A., TSUI, D. C., and GOSSARD, A. C., *Quantized Hall effect at low temperatures*, Phys. Rev. B **25**, 5566 (1982)

PEARSON, G. L. and BARDEEN J., *Electrical Properties of Pure Silicon and Silicon Alloys Containing Boron and Phosphorus*, Physical Review **75**, 865 (1949).

RICHARD, S., ANIEL, F., and FISHMAN, G., *Energy-band structure of Ge, Si, and GaAs: A thirty-band k·p method*, Physical Review B **70**, 235204 (2004).

* SINGH, J. – *Quantum Mechanics: Fundamentals & Applications to Technology.* Weinheim: Wiley VCH, 2004. ISBN 0471157589.

This book on quantum mechanics is written by a semiconductor expert from the University of Michigan. The book is therefore full of application examples in semiconductor physics, which makes it a particularly useful complementary work.

* SINGLETON, J. – *Band theory and the electronic properties of solids.* Oxford: Oxford University Press, 2001. ISBN: 0198506449.

This short book from the Oxford Master Series in Condensed Matter Physics gives a particularly succinct and clear introductory presentation of band theory. It was therefore used as the main source for the revision of this topic, as well as for the discussion of the quantum Hall effect at the end of this book.

SHOCKLEY, W., and READ, Jr., W. T. – *Statistics of the Recombinations of Holes and Electrons*, Physical Review **87**, 835 (1952).

STURGE, M. D. – *Optical Absorption of Gallium Arsenide between 0.6 and 2.75 eV*, Physical Review **127**, 768 (1962).

TEMPLE, P. A. and HATHAWAY, C. E., *Multiphonon Raman Spectrum of Silicon*, Physical Review B **7**, 3685 (1973).

* VASILEVSKIY, M. I. and FERREIRA, I. C. – *Física dos semicondutores: fundamentos, aplicações e nanoestruturas.* Coimbra: Almedina, 2005. ISBN 972402654X.

This is a pioneering work, in Portugal and in Portuguese, with considerable overlap with the contents of this subject.

VILÃO, R. C. – Semiconductors and dielectrics. In *Muon Spectroscopy: an introduction.* (Eds. BLUNDEL, S. J., DE RENZI, R., LANCASTER, T., PRATT, F. L.). Oxford: Oxford University Press, 2022. ISBN 978019885959. p. 157-169.

WALTER, J. P. and COHEN, M. L. – *Calculation of the Reflectivity, Modulated Reflectivity, and Band Structure of GaAs, GaP, ZnSe, and ZnS*, Physical Review **183**, 763 (1969).

WEIJIE, H., YOUWEN, Z., WENRONG, S., MANLONG, D., ZHIYUAN, D. and JUN, Y. – *Residual impurities and electrical properties of undoped LEC InAs single crystals*, Journal of Semiconductors **31**, 042001 (2010).

* YU, Peter Y. and CARDONA, Manuel *Fundamentals of Semiconductors: Physics and Materials Properties*, 2nd edition. Berlin; Heidelberg; New York: Springer-Verlag, 1999. ISBN 354065352X.

This is an advanced book in the physics of semiconductors. It is recommended for those looking for a deeper perspective. This is a reputed advanced classical work which, despite the accelerated pace in this topic, can and should be read/studied by any beginning researcher in this field. It is however a challenging choice for the beginner.

Index

For Product Safety Concerns and Information please contact our EU
representative GPSR@taylorandfrancis.com
Taylor & Francis Verlag GmbH, Kaufingerstraße 24, 80331 München, Germany

www.ingramcontent.com/pod-product-compliance
Ingram Content Group UK Ltd.
Pitfield, Milton Keynes, MK11 3LW, UK
UKHW020205200926
475702UK00004B/623
* 9 7 8 1 0 3 2 9 6 0 5 1 7 *